Measurement Problems in Criminal Justice Research

WORKSHOP SUMMARY

John V. Pepper and Carol V. Petrie

Committee on Law and Justice
and
Committee on National Statistics

Division of Behavioral and Social Sciences and Education

NATIONAL RESEARCH COUNCIL
OF THE NATIONAL ACADEMIES

The National Academies Press
Washington, D.C.
www.nap.edu

THE NATIONAL ACADEMIES PRESS   500 Fifth Street, N.W.   Washington, DC 20001

NOTICE: The project that is the subject of this report was approved by the Governing Board of the National Research Council, whose members are drawn from the councils of the National Academy of Sciences, the National Academy of Engineering, and the Institute of Medicine. The members of the committee responsible for the report were chosen for their special competences and with regard for appropriate balance.

This study was supported by Contract/Grant No. 98-IJ-CX-0030 between the National Academy of Sciences and the National Institute of Justice, Office of Justice Programs, U.S. Department of Justice. Any opinions, findings, conclusions, or recommendations expressed in this publication are those of the author(s) and do not necessarily reflect the views of the organizations or agencies that provided support for the project.

International Standard Book Number 0-309-08635-3

Library of Congress Catalog Card Number 2002115675

Additional copies of this report are available from National Academies Press, 500 Fifth Street, N.W., Lockbox 285, Washington, DC 20055; (800) 624-6242 or (202) 334-3313 (in the Washington metropolitan area); Internet, http://www.nap.edu

Printed in the United States of America

Copyright 2003 by the National Academy of Sciences. All rights reserved.

Suggested citation: National Research Council. (2003). *Measurement Issues in Criminal Justice Research: Workshop Summary*. J.V. Pepper and C.V. Petrie. Committee on Law and Justice and Committee on National Statistics, Division of Behavioral and Social Sciences and Education. Washington, DC: The National Academies Press.

THE NATIONAL ACADEMIES
Advisers to the Nation on Science, Engineering, and Medicine

The **National Academy of Sciences** is a private, nonprofit, self-perpetuating society of distinguished scholars engaged in scientific and engineering research, dedicated to the furtherance of science and technology and to their use for the general welfare. Upon the authority of the charter granted to it by the Congress in 1863, the Academy has a mandate that requires it to advise the federal government on scientific and technical matters. Dr. Bruce M. Alberts is president of the National Academy of Sciences.

The **National Academy of Engineering** was established in 1964, under the charter of the National Academy of Sciences, as a parallel organization of outstanding engineers. It is autonomous in its administration and in the selection of its members, sharing with the National Academy of Sciences the responsibility for advising the federal government. The National Academy of Engineering also sponsors engineering programs aimed at meeting national needs, encourages education and research, and recognizes the superior achievements of engineers. Dr. Wm. A. Wulf is president of the National Academy of Engineering.

The **Institute of Medicine** was established in 1970 by the National Academy of Sciences to secure the services of eminent members of appropriate professions in the examination of policy matters pertaining to the health of the public. The Institute acts under the responsibility given to the National Academy of Sciences by its congressional charter to be an adviser to the federal government and, upon its own initiative, to identify issues of medical care, research, and education. Dr. Harvey V. Fineberg is president of the Institute of Medicine.

The **National Research Council** was organized by the National Academy of Sciences in 1916 to associate the broad community of science and technology with the Academy's purposes of furthering knowledge and advising the federal government. Functioning in accordance with general policies determined by the Academy, the Council has become the principal operating agency of both the National Academy of Sciences and the National Academy of Engineering in providing services to the government, the public, and the scientific and engineering communities. The Council is administered jointly by both Academies and the Institute of Medicine. Dr. Bruce M. Alberts and Dr. Wm. A. Wulf are chair and vice chair, respectively, of the National Research Council.

www.national-academies.org

COMMITTEE ON LAW AND JUSTICE

Charles F. Wellford (*Chair*), Center for Applied Policy Studies and Department of Criminology and Criminal Justice, University of Maryland

Alfred Blumstein, H. John Heinz III School of Public Policy and Management, Carnegie Mellon University

Jeanette Covington, Department of Sociology, Rutgers, The State University of New Jersey

Ruth Davis, The Pymatuning Group, Inc., Alexandria, VA

Jeffrey Fagan, Schools of Law and Public Health, Columbia University

Darnell Hawkins, Department of African American Studies, University of Illinois, Chicago

Philip Heymann, Center for Criminal Justice, Harvard Law School

Candace Kruttschnitt, Department of Sociology, University of Minnesota

Mark Lipsey, Department of Psychology and Human Development, Vanderbilt University

Colin Loftin, School of Criminal Justice, State University of New York at Albany

John Monahan, School of Law, University of Virginia

Daniel Nagin, H. John Heinz III School of Public Policy and Management, Carnegie Mellon University

Joan Petersilia, School of Social Ecology, University of California, Irvine

Peter Reuter, Department of Criminology and Research, University of Maryland

Wesley Skogan, Department of Political Science and Institute for Policy Research, Northwestern University

Cathy Spatz Widom, Department of Psychiatry, New Jersey Medical School

Kate Stith, School of Law, Yale University

Michael Tonry, Institute of Criminology, Cambridge University

Carol Petrie, *Director*
Ralph Patterson, *Senior Project Assistant*

COMMITTEE ON NATIONAL STATISTICS

John E. Rolph (*Chair*), Marshall School of Business, University of Southern California
Joseph G. Altonji, Department of Economics, Northwestern University
Lawrence D. Brown, Department of Statistics, University of Pennsylvania
Julie DaVanzo, RAND, Santa Monica, CA
William F. Eddy, Department of Statistics, Carnegie Mellon University
Robert M. Groves, Joint Program in Survey Methodology, University of Maryland, College Park
Hermann Habermann, Statistics Division, United Nations, New York
Joel Horowitz, Department of Economics, University of Iowa
William D. Kalsbeek, Department of Biostatistics, University of North Carolina, Chapel Hill
Roderick J.A. Little, School of Public Health, University of Michigan
Thomas A. Louis, RAND, Arlington, VA
Daryl Pregibon, AT&T Laboratories-Research, Florham Park, NJ
Francisco J. Samaniego, Division of Statistics, University of California, Davis
Richard L. Schmalensee, Sloan School of Management, Massachusetts Institute of Technology
Matthew D. Shapiro, Department of Economics, University of Michigan

Andrew A. White, *Director*

Acknowledgments

Accurate, reliable, and valid measurement of crime and criminal victimization is becoming increasingly important. American anxiety about violent crime remains high even when rates are plummeting. However, rates of violent crime are again on the rise, and as advances in technology continue to make the world ever smaller, new types of crime are emerging. This workshop grew out of a need to develop better information for policy officials and researchers about crime in the United States—how much crime takes place in general, how best to develop estimates of crime in small geographic areas and among subpopulations, and how to estimate the frequency of rare but heinous crimes. In general, workshop participants focused on how scientific advances in research and analysis can be used to improve the design of surveys. Participants joked that only about 200 people in the world really worry about these matters, but they are important because without accurate and reliable estimates of crime we have no hope of understanding what works or does not to make society safer.

The Committee on Law and Justice and the Committee on National Statistics were fortunate to be able to bring together for this workshop many of the scholars who care the most about these issues. Four papers were commissioned. Two are published here as articles signed by the authors. The others, in the interest of time and space, are summarized, with the discussion, by the authors of this report. Many people made generous contributions to the workshop's success. We thank the authors of the papers presented—Roger Tourangeau and Madeline E. McNeeley, University of Maryland; Terence P. Thornberry and Marvin D. Krohn, University at Albany, State University of New York; Trivellore Raghunathan, Institute for Social Research, University of Michigan; and Richard McCleary, Douglas Weibe, and David Turbow, University of California, Irvine. We also thank the scholars who prepared comments for each of the papers: Alfred Blumstein, Carnegie Mellon University; Laura Dugan, Georgia State University; David Farrington, Cambridge University; Judith Lessler, Research Triangle Institute; James Lynch, American University; Charles Manski, Northwestern University; Elisabeth Stasny, Ohio State University; and James Walker, University of Wisconsin, Madison.

The authors are particularly grateful for the leadership of Colin Loftin, University at Albany, State University of New York, who guided the organization of the workshop, made all of the initial contacts with paper authors and many of the discussants to develop workshop themes, ably chaired the workshop sessions, and provided comments and guidance for the development of this report. Special thanks also go to Steven Feinberg, Carnegie Mellon University, for his advice and guidance as the workshop topics were developed and to William Eddy, Robert Groves, and Charles Manski, Committee on National Statistics, and Andrew White, director of the Committee on National Statistics for their invaluable help in shaping the workshop. We also thank Christine McShane for her editorial support and Yvonne Wise for managing the production process.

This workshop summary has been reviewed in draft form by individuals chosen for their diverse perspectives and technical expertise, in accordance with procedures approved by the Report Review Committee of the National Research Council. The purpose of this independent review is to provide candid and critical comments that will assist the institution in making the published report as sound as possible and to ensure that the report meets institutional standards for objectivity, evidence, and responsiveness to the study charge. The review comments and draft manuscript remain confidential to protect the integrity of the deliberative process.

We thank the following individuals for their participation in the review of this report: Laura Dugan, Georgia State University; David Farrington, Cambridge University; Judith Lessler, Research Triangle Institute; James Lynch, The American University; Charles Manski, Northwestern University; Elizabeth Stasny, Ohio State University; and James Walker, University of Wisconsin, Madison.

Although the reviewers listed above provided many constructive comments and suggestions, they were not asked to endorse the content of the report nor did they see the final draft of the report before its release. The review of this report was overseen by Alfred Blumstein, Carnegie Mellon University. Appointed by the National Research Council, he was responsible for making certain that an independent examination of the report was carried out in accordance with institutional procedures and that all review comments were carefully considered. Responsibility for the final content of this report rests entirely with the authoring committee and the institution.

Contents

1 Overview   1
   John V. Pepper and Carol V. Petrie

2 Measuring Crime and Crime Victimization:
   Methodological Issues   10
   Roger Tourangeau and Madeline E. McNeeley

3 Comparison of Self-Report and Official Data for
   Measuring Crime   43
   Terence P. Thornberry and Marvin D. Krohn

Appendixes
A   Workshop Agenda   95

B   List of Workshop Participants   98

1

Overview

John V. Pepper and Carol V. Petrie

It is axiomatic that accurate and valid data and research information on both crime and victimization are critical for an understanding of crime in the United States and for any assessment of the quality of the activities and programs of the criminal justice system. In July 2000 the Committee on Law and Justice and the Committee on National Statistics of the National Research Council convened a workshop to examine an array of measurement issues in the area of crime victimization and offending and to explore possible areas for future research to improve measurement methods. This report provides information that was presented at the workshop.

TWO MAJOR DATA SOURCES

Most measurement of crime in this country emanates from two major data sources. For almost seven decades, the FBI's Uniform Crime Reports (UCR) has collected information on crimes known to the police and arrests from local and state jurisdictions throughout the country. The National Crime Victimization Survey (NCVS), a general population survey designed to discover the extent, nature, and consequences of criminal victimization, has been conducted annually since the early 1970s. Other national surveys that focus on specific problems, such as delinquency, violence against women, and child abuse, also provide important data on crime, victims, and offenders.

These data collection systems utilize different methods of measuring criminal behavior. The UCR relies on official data that have been collected and reported by law enforcement agencies. The NCVS and other surveys discussed in this report are large-scale social surveys that rely on self-reports of offenses or victimization.

Although these data collection systems do many things right, they are, like any such system, beset with the methodological problems of surveys in general as well as particular problems associated with measuring illicit, deviant, and deleterious activities. Such problems include nonreporting and false reporting, nonstandard definitions of events, difficulties associated with asking sensitive questions, sampling problems such as coverage and nonresponse, and an array of other factors involved in conducting surveys of individuals and implementing official data reporting systems.

Compounding these problems are the recent interest in rare crime events, such as violent crimes committed by youth and hate crimes; the need for attention to vulnerable subpopulations, such as very young and school-age children and disabled, elderly, and immigrant populations; and a focus on small- or local-area estimates of crime and victimization. Congress periodically requires the U.S. Department of Justice to develop new research or data collection efforts to measure crime victimization in specific populations and for small areas. Understanding victimization and offending in these subgroups, however, can be particularly difficult.

In general, criminal victimization is a relatively rare event—that is, in any given reference period, the majority of respondents do not report any victimization. Very large general population samples are therefore required to accurately characterize the population of offenders and victims, and detailed subgroup analyses can be problematic. Some important subgroups may not be covered at all (e.g., homeless people), and smaller research studies of crimes against these subgroups often have problems of statistical power because of small sample sizes in most cases. For many hard-to-identify subpopulations, such as people with disabilities and abused children, there is no large, well-defined group from which to draw a sample for measuring victimization—in other words, a sampling frame. This, as well as more conventional problems associated with interviewing crime victims, presents substantial design and analytical difficulties.

Official data such as UCR arrest data have a different set of problems. Foremost among them is that most crimes are not reported to the police, and only a small proportion of those that are reported result in an arrest. Increases or decreases in reports or in arrests for certain offenses, such as burglary or auto theft, can therefore result in large differences in outcomes and misleading conclusions about crime trends.

The accuracy of official data is also compromised by differences in the definitions of crimes and reporting protocols. Most national-level official data are compiled through the voluntary reporting of local-level agencies— 17,000 law enforcement agencies nationwide for arrests; a sample of 300 prosecutors' offices nationwide for prosecution data, for example. However, these agencies do not always file reports as called for in the reporting protocol. A review of the 1999 UCR data posted on the FBI's web site indicates that six states—Illinois, Kansas, Kentucky, Maine, Montana, and New Hampshire—report only limited data. In Illinois, for example, only six cities with populations of 10,000 or more report arrest data. Rape data were unavailable for two states because the state reporting agencies did not follow the national UCR guidelines (available: <http://www.FBI.gov/ucr/99cius.htm> [8/15/01]).

There has been a significant lack of governmental and private interest and investment in research aimed at solving these kinds of problems. It has now been more than 30 years since large-scale data systems on crime victimization have been under way and almost that long since research questions have been included as part of the structure of the NCVS. A major research effort undertaken in the early 1980s as part of the redesign of the NCVS addressed many important sampling questions, but (according to several workshop participants) for social and political reasons, much of this information did not make it into the survey.[1]

Today, in fact, the problems may be growing worse because of eroding federal investment in data systems and social science research on crime and victimization. The sample in the NCVS has shrunk because of flat funding over the past 25 years, from 60,000 households (in 1974) to the current level of 42,000 households (Patsy Klaus, senior statistician, Bureau of Justice Statistics, personal communication, 7/24/02). The promise of improvements in data on reported crimes through conversion to an incident-based reporting system has not been realized because of a lack of funding to

---

[1]Participants noted that many of the papers commissioned to support the redesign of the NCVS are buried in archives or personally held files; as a result, a handful of government officials and scholars know more about how to solve some of these problems than many academics and policy officials realize, but the kind of careful experimentation with different methods, which is required to develop standards against which things can be measured, has not taken place.

support the necessary changes at the state and local levels. Except for modest new funds to study violence against women, the federal budget for social science research on crime and victimization also has remained flat for two decades. Available funds generally are reserved for studies that potentially have a direct impact on policy; support for longitudinal or methodological studies for the most part simply is not there.

SCOPE OF THE REPORT

The workshop was designed to consider similarities and differences in the methodological problems encountered by the survey and criminal justice research communities and, given the paucity of available funding, to consider on what the academic community might best focus its research efforts. Participants represented a range of interests and academic disciplines, including criminology, econometrics, law, psychology, public policy, and statistics. Representatives of two federal agencies, the Bureau of Justice Statistics and the National Institute of Justice, also participated. The Workshop Agenda (Appendix A) and List of Workshop Participants (Appendix B) appear at the end of this workshop summary.

In addition to comparing and contrasting the methodological issues associated with self-report surveys and official records, the workshop explored methods for obtaining accurate self-reports on sensitive questions about stigmatized (crime) events, estimating crime and victimization in rural counties and townships, and developing unbiased prevalence and incidence rates for rare events among population subgroups. There also was a discussion of how to define and measure new types of crime: street crimes, such as carjacking, which combines car theft with robbery, assault, and sometimes homicide; and cyber crime, which combines ordinary forms of white-collar crime, such as fraud, with technology-based offenses, such as hacking.

The discussions prompted participants to think about the importance of good crime statistics, which according to some participants have never gotten much beyond measures of prevalence and incidence. How might more accurate figures serve the public interest? These data collection systems and measures are routinely used to talk about crime trends and the effect of interventions on crime rates and specific crime types, but because of what is known about the errors in them and what is tacitly assumed about the unknown errors—that they are even greater—researchers generally seem uncomfortable about using the available data to draw conclusions about the scope and nature of crime problems. Participants discussed the need to prioritize these measurement problems, to think carefully about which ones are most important, and to focus on the two or three of primary concern.

To facilitate the workshop discussion, four papers were commissioned in advance, and they form the basis of this report. Following the workshop, the authors were invited to revise their papers in response to the workshop discussion.

Two chapters follow this introduction. Chapter 2, a paper by Roger Tourangeau and Madeline E. McNeeley, is on methodological issues in measuring crime and crime victimization. One discussant summarized this paper as an excellent exposition on why context matters. Chapter 3, a paper by Terence P. Thornberry and Marvin D. Krohn, is on the strengths and weaknesses of the self-report method in studies of offenders, and how to make improvements. A brief summary of the topics addressed in the other two commissioned papers appears below. It discusses particular issues that can arise when using surveys to draw inferences on specific subgroups, and methodological issues associated with measuring crimes in small geographic areas, from papers respectively by Richard McCleary, Douglas Wiebe, and David Turbow and by T.E. Raghunathan.

SUBGROUP AND SMALL-AREA ESTIMATION OF CRIME AND VICTIMIZATION

Public attention to crime and victimization often focuses on particular subgroups in which deviant behavior may be most troublesome. For example, hate crimes, crimes committed by youth, and crimes committed against vulnerable subpopulations including children, the elderly, and people with disabilities have all been the subject of recent investigations and legislation. Similarly, there has been increased focus in local-area criminal activity.

Interest in evaluating small subgroups of the population, however, results in especially difficult data and methodological problems. As noted above, crime and victimization are rare events. Thus, drawing precise conclusions about small subgroups requires that careful attention be paid to the sampling frame and the methods required to link the data to the subgroups of interest. Workshop participants spent some time discussing these issues, stressing the need for both reliable data and credible assumptions. Particular attention was focused on well-known but difficult problems for drawing inferences on selected subpopulations, screening bias, and small-area estimation.

Screening Biases

"Screening" refers to the basic survey methods used to identify cases or subgroups of interest (Morrison, 1985; Thorner and Remein, 1961; U.S. Commission on Chronic Illness, 1957). While the basic idea is both simple and appealing, screening for low-incidence events or cases can lead to substantial and systematic misreporting errors.

In Chapter 2, for example, screening errors are discussed in the context of false positive reports of defensive gun uses. A second example, which was explored in detail by McCleary and colleagues, centers on the victimization of persons with disabilities, and particularly persons with cognitive or developmental disabilities, and the extent to which a high percentage of false positives—persons identifying themselves as disabled when they are not—from general population samples can distort any estimates of the correlates of that particular category.[2]

False positive reports of rare events or subpopulations of interest can lead to substantial biases. In their paper, McCleary, Wiebe, and Turbow also demonstrated the potential biases created by screening respondents to select out rare subgroups of interest. These biases, the authors argue, should be sufficient to rule out any attempt to estimate low-probability events with a sample survey screening design, such as the National Crime Victimization Survey, and to instead develop estimates using a sampling frame that targets the specific group of interest. The authors also recommend using state-of-the-art survey methodologies that apply what is known about cognitive processes to interviewing techniques (also discussed by Tourangeau and McNeeley in Chapter 2).

Small-Area Estimates

For many research and policy questions, it is important to analyze crime data at a relatively low level of aggregation, such as county or even census tract. The problem, however, is that many of the national surveys, such as the National Victimization Survey, though large enough to yield national

---

[2]Public Law 105-301 (105th Congress, October 27, 1998) mandates the federal government to collect victimization data on disabled persons.

estimates, or even state-level estimates, with adequate precision, are inadequate to yield reliable estimates for small areas. The raw rates are unstable, due to small numerators and/or denominators. Unstable rates are problematic because a change of only a few events in the numerator can result in large changes in rates. Furthermore, extreme rates (high or low) are often the result of inherent variability (noise) rather than true extremes in the phenomenon. The areas with the smallest populations will have the extreme values and will dominate a map or a statistical analysis. The least precise rates thus have the most influence.

In his paper, T.E. Raghunathan developed a framework for obtaining small or local-area estimates of victimization or crime rates by combining information from surveys and administrative data sources. Using these data, the measures created from the administrative data sources can be treated as predictors and the victimization rates computed from the surveys as dependent variables. This approach depends on the availability of a good set of predictors—that is, detailed data—and the credibility of the assumptions used in the prediction model.

PROGRESS AND CONTINUED PROBLEMS

As the workshop participants acknowledged, tremendous progress has been made in the past 50 years in our ability to measure, describe, and evaluate both victimization and criminal behavior. The development and refinement of a number of administrative and survey data systems have had and continue to play a central role in the ability of the nation to understand crime. However, many difficult barriers remain to drawing credible inferences using these data systems. The classic survey sampling problems of response and coverage are of special concern when attempting to measure stigmatized and other deleterious events. The transitory nature of the different types of criminal activity and of the interests of policy makers presents unique problems for survey designers and researchers. Likewise, continued interest in evaluating specific subpopulations and geographic areas requires improved data and research methodologies.

Workshop participants recognized the need to be pragmatic. The federal budget for data on crime and justice is modest and has not grown in nearly two decades. Without additional resources devoted to improving the data on crime and justice, many of the problems discussed at the workshop are likely to persist. In fact, to some degree, all of the issues discussed have been studied and described for over two decades without resolution.

The participants focused on the importance of two overriding issues: (1) there are significant and substantive measurement problems with the existing surveys that are likely to remain unresolved without additional funding support; (2) priorities need to be established so that the available resources devoted to these issues can be used most effectively.

In addition to discussions of these general points, much of the workshop focused on measurement problems that arise from the existing surveys and, to some degree, on future directions for research to resolve them. The measurement issues discussed include, but certainly are not limited to, the following:

- *Improving reliability.* Many participants suggested that additional attention should be focused on improving the reliability and validity of self-report surveys, rather than simply assessing these characteristics. For example, Thornberry and Krohn (Chapter 3) argue that "it is likely that both validity and reliability would be improved if we experimented with alternative items for measuring the same behavior and identified the strongest ones." McCleary, Wiebe, and Turbow suggest using multiple data sources and questions on the same item to "develop an estimate of the prevalence and magnitude of the bias for an item and adjust for it statistically."

- *Evaluating the impact of nontraditional survey methodologies.* Much of the knowledge about reporting errors in crime surveys comes from cross-sectional studies. Using self-reports in longitudinal studies, especially ones that cover major portions of the life course, creates a new set of challenges. Likewise, Thornberry and Krohn argue for the need to better understand the effects of self-administration (e.g., computer-assisted self-administered interviews) on reporting errors in surveys of crime.

- *Response errors across self-report and administrative surveys.* Many participants argued for a more systematic analysis comparing and contrasting the understanding of crime and victimization in self-report and official data.

- *Survey packaging design.* Tourangeau and McNeeley (Chapter 2) wonder whether "the apparent topic of the survey, the survey's sponsorship, the organization responsible for collecting the data, the letterhead used on advance letters, and similar procedural details will affect respondents' views about the need for accuracy in reporting and about the type of incidents they are supposed to report."

In addition to focusing on improving the validity and reliability of surveys and administrative data, there was some discussion at the workshop about the need to develop and apply credible models. Even perfectly valid and reliable data cannot completely address the many questions of interest. This was particularly obvious in Dr. Raghunathan's presentation on developing small-area estimates and in the more general discussion that followed. Evaluations of subgroups or geographic areas that may not be adequately represented or covered by the sampling scheme invariably require researchers to make assumptions. The credibility of empirical findings, however, depends on the validity of the maintained assumptions. Some participants expressed concern that little thought was being paid to the assumptions used to derive inferences on crime and victimization, especially for those regarding specialized populations or geographic areas.

The variety of ideas, concerns, and recommendations raised by the papers and commentary at the workshop resulted in a rich discussion of crime measurement and research issues. Public opinion polls have repeatedly demonstrated that crime is a policy issue of pre-eminent concern to the American public. The goal of the Committee on Law and Justice and the Committee on National Statistics in convening the workshop, commissioning the papers, and issuing this report is to stimulate further discussion and eventually a greater focus on the importance of improving data collection systems and measurement of crime.

REFERENCES

McCleary, R., D. Wiebe, and R. Turbow
    2000 Screening Bias. Paper commissioned for the Committee on Law and Justice, Workshop on Measurement Problems in Criminal Justice Research, July 2000, National Research Council, Washington, DC.

Morrison, A.S.
    1985 *Screening in Chronic Disease.* Cary, NC: Oxford University Press.

Raghunathan, T.E.
    2000 Combining Information from Multiple Sources for Small Area Estimation of Victimization Rates. Paper commissioned for the Committee on Law and Justice, Workshop on Measurement Problems in Criminal Justice Research, July 2000, National Research Council, Washington, DC.

Thorner, R.M., and Q.R. Remein
    1961 Principles and Procedure in the Evaluation of Screening for Disease. *Public Health Monograph #67.* Washington, DC: U.S. Government Printing Office.

U.S. Commission on Chronic Illness
    1957 *Chronic Illness in the United States.* Cambridge: Harvard University Press.

2

Measuring Crime and Crime Victimization: Methodological Issues

Roger Tourangeau and Madeline E. McNeeley

All surveys face measurement challenges, but few topics raise problems of the variety or seriousness of those involved in measuring crime and crime victimization. As Skogan (1981) points out in his thoughtful monograph, *Issues in the Measurement of Victimization*, the nature of crime and crime victimization adds wrinkles to virtually every standard source of error in surveys. For example, even in our relatively crime-ridden times, crimes remain a rare event and, as a result, survey estimates are subject to large sampling errors. One national survey (National Victims Center, 1992) estimated that 0.7 percent of American women had experienced a completed rape during the prior year. This estimate was based on a sample of 3,220 responding women, implying that the estimate reflected positive answers to the relevant survey items from about 23 women. Skogan details the large margins of sampling error for many key estimates from the National Crime Survey (NCS), a survey that used a very large sample (and which later evolved into the National Crime Victimization Survey).

The clandestine nature of many crimes means that the victim may be unable to provide key details about the victimization and may not even be aware that a crime has been committed at all. Certain incidents that are supposed to be reported in an interview may seem irrelevant to respondents, since they do not think of these incidents as involving *crimes*. For example, victims of domestic violence or of sexual harassment may not think of these as discrete criminal incidents but as chronic family or interpersonal problems. It may be difficult to prompt the recall of such incidents with the short, concrete items typically used in surveys (for a fuller discussion, see Skogan, 1981:7-10).

NATIONAL CRIME VICTIMIZATION SURVEY

The National Crime Survey and its successor, the National Crime Victimization Survey (NCVS), underwent lengthy development periods featuring record check studies and split-ballot experiments to determine the best way to measure crime victimization. In the records check studies, the samples included known crime victims selected from police records. In survey parlance, these were studies of reverse records check—the records had been "checked" before the survey reports were ever elicited. The studies were done in Washington, D.C., Akron, Cleveland, Dayton, San Jose, and Baltimore (see Lehnen and Skogan, 1981, for a summary). A key objective of these early studies was to determine the best length for the reporting period for a survey, balancing the need to increase the number of crime reports with the need to reduce memory errors.

A second wave of studies informing the NCS design was carried out in the early 1980s by researchers at the Bureau of Social Science Research and the Survey Research Center at the University of Michigan (summarized by Martin et al., 1986). This second wave of developmental studies mainly involved split-ballot comparisons (in which random portions of the sample were assigned to different versions of the questionnaire) focusing on the "screening" items, in which respondents first indicate they have relevant incidents to report. Some of these studies were inspired by a conference (described in Biderman, 1980) that brought cognitive psychologists and survey researchers together to examine the memory issues raised by the NCS. Unfortunately, some of the most intriguing findings from the resulting experiments were never published and are buried in hard-to-find memoranda.

For several reasons, the NCVS results are widely used as benchmarks to which statistics from other surveys on crime and crime victimization are compared. Conducted by the Bureau of the Census, the NCVS is the largest and oldest of the crime victimization studies. It uses a rotating panel design in which respondents are interviewed several times before they are "retired" from the sample, a design that greatly improves the precision of sample estimates. It uses a relatively short, six-month reporting period and "bounded" interviewing, in which respondents are instructed to report only incidents that have occurred since the previous interview and are reminded of the incidents they reported then. (Results of the first interview, which is necessarily unbounded, are discarded.) The initial interview is done face to face to ensure maximum coverage of the population; if necessary, subsequent interviews are also conducted in person.

Examples of Measurement Problems

Despite these impressive design features and the large body of methodological work that shaped it, the NCVS is not without its critics. Two recent controversies illustrate the problems of the NCVS and of crime surveys more generally. One controversy centers on the number of incidents of defensive gun use in the United States; the other concerns the number of incidents of rape. In both cases, seemingly similar surveys yield widely discrepant results; the ensuing methodological controversies point to unresolved issues in how to collect data on crime, gun use, and crime victimization in surveys.

Defensive Gun Use

In 1994, McDowall and Wiersema published an estimate of the number of incidents over a four-year period in which potential crime victims had used guns to protect themselves during an actual or attempted crime. Their estimate was based on data from the NCVS, which gathers information about several classes of crime—rape, assault, burglary, personal and household larceny, and car theft. When respondents report an incident in which they were victimized, they are asked several follow-up questions, including, "Was there anything you did or tried to do about the incident while it was going on?" and, "Did you do anything (else) with the idea of protecting yourself or your property while the incident was going on?" Responses to these follow-up probes are coded into a number of categories, several of which capture defensive gun use (e.g., "attacked offender with gun"). The key estimates McDowall and Wiersema presented were that between 1987 and 1990 there were some 260,000 incidents of defensive gun use in the United States, roughly 65,000 per year. Although big numbers, they pale by comparison with the total number of crimes reported during the same period—guns were used defensively in fewer than one in 500 victimizations reported in the NCVS; moreover, criminal offenders were armed about 10 times more often than their victims. These are just the sort of statistics dear to gun control advocates. McDowall and

Wiersema conclude that "criminals face little threat from armed citizens" (1994:1984).

McDowall and Wiersema note, however, that their estimates of defensive gun use differ markedly from those based on an earlier survey by Kleck (1991; see also Kleck and Gertz, 1995). Kleck's results indicated 800,000 to 1 million incidents of defensive gun use annually. These numbers were derived from a national telephone survey of 1,228 registered voters who were asked: "Within the past five years have you, yourself, or another member of your household used a handgun, even if it was not fired, for self-protection or for the protection of property at home, work, or elsewhere, excluding military service and police security work?"

There are so many differences between the Kleck survey and the NCVS that it should come as no surprise that the results do not line up very well. The two surveys covered different populations (the civilian noninstitutional population in the NCVS versus registered voters with a telephone in the Kleck survey), interviewed respondents by different methods (in-person versus telephone), covered different recall periods (six months in NCVS versus five years in the Kleck study), and asked their respondents markedly different questions. The NCVS uses a bounded interview, the Kleck survey an unbounded interview. Still, the difference between 65,000 incidents a year and some 900,000 is quite dramatic and would seem to demand a less mundane explanation than one involving routine methodological differences. A later telephone survey by Kleck and Gertz (1995) yielded an even higher estimate—2.5 million incidents of defensive gun use.

McDowall and Wiersema (1994) cite two other possible explanations of the differences between the results of the NCVS and the earlier Kleck study. First, they note that Kleck's estimates rest on the reports of a mere 49 respondents. (The later Kleck and Gertz estimates also rest on a similarly small base of positive reports—66 out of nearly 5,000 completed interviews.) Even a few mistaken respondents could have a large impact on the results. In addition, the Kleck item covers a much broader range of situations than does the NCVS. The NCVS excludes preemptive use of firearms (e.g., motorists who keep a gun in their car "for protection" but never take it out of the glove compartment), focusing more narrowly on gun use during actual or attempted crimes. It is possible that much of the disparity between the NCVS estimates and those derived from the two Kleck studies reflects the broader net cast in the latter surveys.

A methodological experiment by McDowall, Loftin, and Presser (2000) compared questions modeled on the ones used in the NCVS with ones like those used in the Kleck surveys. Respondents were asked both sets of questions—both written to cover a one-year recall period—and the experiment varied which ones came first in the interview. The sample included 3,006 respondents, selected from a list of likely gun owners. Overall, the Kleck items yielded three times more reports of defensive gun use than the NCVS-style items. What was particularly interesting in the results was that the two sets of items appeared to yield virtually nonoverlapping sets of incidents; of the 89 reports of defensive gun use, only 9 were mentioned in response to both sets of items.

Prevalence of Rape

An even more disparate set of figures surrounds the issue of the number of women in the United States who have been the victim of attempted or completed rapes. Once again, the studies from which the estimates are drawn differ in many crucial particulars—they sample different populations, ask different questions that are based on different definitions of rape, conduct data collection via different methods, and cover different recall periods. As with the estimates of defensive gun use, what is surprising is not that the estimates differ from each other but that they differ so widely.

Several studies converge on the estimate that about one-quarter of American women have been victims of completed or attempted rape at some time in their lives (see, for example, Koss, 1993:Table 1). Most of these figures do not accord well with the rape estimates from the NCVS; the NCVS covers a more limited period—six months—and does not produce estimates of lifetime victimization. But the NCVS's annual estimates—for example, fewer than 1 woman or girl in 1,000 experienced a rape or attempted rape in 1992 (see Koss, 1996)—imply that rape is much less common than indicated by most of the other surveys. Koss (1992, 1993, 1996) has been an energetic critic of the NCVS procedures for assessing the prevalence of rape, but at least two other papers have presented careful comparisons between the NCVS procedures and those of other surveys (Fisher and Cullen, 2000; Lynch, 1996) and support Koss's contention that methodological details matter a great deal in assessing the prevalence of rape victimization.

Lynch, for example, reports about a twofold difference between annual estimates for 1992 from the NCVS and the National Women's Study (NWS) for the previous year. For 1992 the NCVS estimated 355,000 incidents of rape; by contrast, the NWS estimated that 680,000 women were rape victims. (The NCVS figure translates into fewer than 355,000 victims since the same person may have experienced multiple victimizations.) Lynch explores a number of differences between the two studies, including:

- the age range covered by the surveys (18 and older for the NWS, 12 and older for the NCVS);
- the sample sizes (4,008 for the NWS, 100,000+ for the NCVS);
- the length of the recall period (one year for the NWS, six months for the NCVS);
- the schedule of interviewing (the NWS estimates are based on data from the second interview from a three-wave longitudinal study, whereas the NCVS estimates reflect data from the second through last interviews from a seven-wave panel); and
- the questions used (brief yes-no items in the NWS versus detailed incident reports in the NCVS).

Despite the methodological differences between the two surveys, the difference between the two estimates is probably not significant. The estimates from both surveys have large standard errors (approximately 190,000 for the NWS estimate and approximately 32,000 for the NCVS), and the standard error of the difference is on the order of 200,000.

One major difference between the NCVS and most of the other surveys assessing the frequency of rape involves the basic strategy used to elicit reports about rapes and other crimes. The NCVS begins with a battery of yes-no items designed to prompt reports about a broad array of completed or attempted crimes. Only one of these initial screening items directly mentions rape ("Has anyone attacked or threatened you in any of these ways . . . any rape, attempted rape, or other type of sexual attack?"), although several other questions concern actual or threatened violence. Once the respondent completes these initial screening items, further questions gather more detailed information about each incident; the final classification of an incident in the NCVS reflects these detailed reports rather than the answers to the initial screening questions. Most of the other surveys on rape differ from this procedure in two key ways—first, they ask multiple screening questions specifically crafted to elicit reports about rape and, second, they omit the detailed follow-up questions. For example, a survey by Koss, Gidycz, and Wisniewski (1987) included five items designed to elicit reports of attempted or completed rape. The items are quite specific. For example, one asks, "Have you had a man attempt sexual intercourse (get on top of you, attempt to insert his penis) when you didn't want to by threatening or using some degree of force (twisting your arm, holding you down, etc.) but intercourse did not occur?" The NWS adopted this same approach, employing five quite explicit items to elicit reports of attempted or completed rape.

There is little doubt that including multiple concrete items will clarify the exact concepts involved and prompt fuller recall. Multiple items provide more memory cues and probably trigger more attempts at retrieval; both the added cues and the added time on task are likely to improve recall (Bradburn and Sudman, 1979; Burton and Blair, 1991; Cannell et al., 1981; Means et al., 1994; Wagenaar, 1986; Williams and Hollan, 1981). The NCVS is a general-purpose crime survey, and its probes cover a broad array of crimes. The NWS and the Koss surveys use much more detailed probes that focus on a narrower range of crimes. At the same time, the absence of detailed information about each incident could easily lead to classification errors.

A study by Fisher and Cullen (2000) included both yes-no screening items of the type used by Koss and colleagues, the NWS, and many other studies of rape and the more detailed questions about each incident featured by the NCVS. They compared responses to the screening questions with the final classifications of the incidents based on the detailed reports. There were twice as many positive answers to the rape screening questions as there were incidents ultimately classified as rapes based on the detailed reports. (The rape screening items also captured many incidents involving some other type of sexual victimization.) In addition, some incidents classified as rapes on the basis of the detailed information were initially elicited by screening items designed to tap other forms of sexual victimization. The results suggest that, even when the wording of screening items is quite explicit, respondents can still misclassify incidents.

Factors Affecting Reporting in Crime Surveys

Many surveys on sensitive subjects adopt methods primarily designed to reduce underreporting—that is, the omission of events that should, in principle, be reported. And it is certainly plausible that women would be reluctant to report extremely painful and personal incidents such as attempted or completed rapes. Even with less sensitive topics, such as burglary or car theft, a variety of processes—lack of awareness that a crime has been committed, forgetting, unwillingness to work hard at answering—can lead to systematic underreporting. There are also reasons to believe that crime surveys, like other surveys that depend on recall, may be prone to errors in the opposite direction as well. Because crime is a relatively rare event, most respondents are not in the position to omit eligible incidents; they do not have any to report. The vast majority of respondents can only overreport defensive gun use, rapes, or crime victimization more generally.

In his discussion of the controversy over estimates of defensive gun use, Hemenway (1997) makes the same point. All survey questions are prone to errors, including essentially random reporting errors. For the moment, let us accept the view that 1 percent of all adults used a gun to defend themselves against a crime over the past year. If the sample accurately reflects this underlying distribution, then only 1 percent of respondents are in the position to underreport defensive gun use; the remaining 99 percent can only overreport it. Even if we suppose that an underreport is, say, 10 times more likely than an overreport, the overwhelming majority of errors will still be in the direction of overreporting. If, for example, one out of every four respondents who actually used a gun to defend himself denies it while only 1 in 40 respondents who did not use a gun in self-defense claim in error to have done so, the resulting estimate will nonetheless be sharply biased upward (1% × 75% + 99% × 2.5% = 3.25%). It is not hard to imagine an error rate of the magnitude of 1 in 40 arising from respondent inattention, misunderstanding of the questions, interviewer errors in recording the answers, and other essentially random factors. Even the simplest survey items—for instance, those asking about sex and age—yield less than perfectly reliable answers. Random errors can, in the aggregate, yield systematic biases when most of the respondents are in the position to make errors in only one direction.

Aside from sheer unreliability, though, reporting in crime surveys may be affected by several systematic factors that can introduce additional distortions of their own. We focus on two of these systematic factors here. First, we address the potentially sensitive nature of the questions on many crime surveys and the impact of the mode of data collection on the answers to such questions. This is followed by an examination of the effects of the context in which survey items are presented, including the physical setting of the interview, the perceived purpose and sponsorship of the study, and prior questions in the interview.

IMPACT OF THE MODE OF DATA COLLECTION

Most of the surveys that have produced the widely varying estimates of defensive gun use and rape incidence use some form of interviewer administration of the questions. For example, Koss (1993) lists 20 surveys on sexual victimization of women; only 4 (all of them involving local samples from individual communities) appear to use self-administered questionnaires. The remainder rely on interviewers to collect the data, in either face-to-face or telephone interviews. (The NCVS uses both methods; the initial interview is done face to face, but later interviews are, to the extent possible, done by telephone.) The last decade has seen dramatic changes in the methods used to collect survey data, including the introduction of several new methods of computerized self-administration. For example, the National Household Survey of Drug Abuse, a large survey sponsored by the Substance Abuse and Mental Health Services Administration, has adopted audio computer-assisted self-interviewing (ACASI). With ACASI a computer simultaneously displays the item on screen and plays a recording of it to the survey respondent via earphones. The respondent enters an answer directly into the computer using the keypad. Other new methods for administering questions include computer-assisted self-interviewing without the audio (CASI), e-mail surveys, telephone ACASI, and World Wide Web surveys.

Two trends have spurred the development and rapid adoption of new methods of computerized self-administration of surveys. First, various technological changes—such as the introduction of lighter, more powerful laptop computers, development of the World Wide Web, widespread adoption of e-mail, and improvements in sound card technology—have made the new methods possible. Second, the need for survey data on sensitive topics, such as illicit drug use and sexual behaviors related to the spread of AIDS, has made the new methods highly desirable, since they combine the privacy of self-administration with the power and flexibility of computer administration. Widespread interest in the new methods has spurred survey methodologists to reexamine the value of self-administration for collecting survey data on sensitive topics.

Gains from Self-Administration

There is strong evidence to support the value of self-administration for eliciting reports about sensitive behaviors. To illustrate the gains from self-

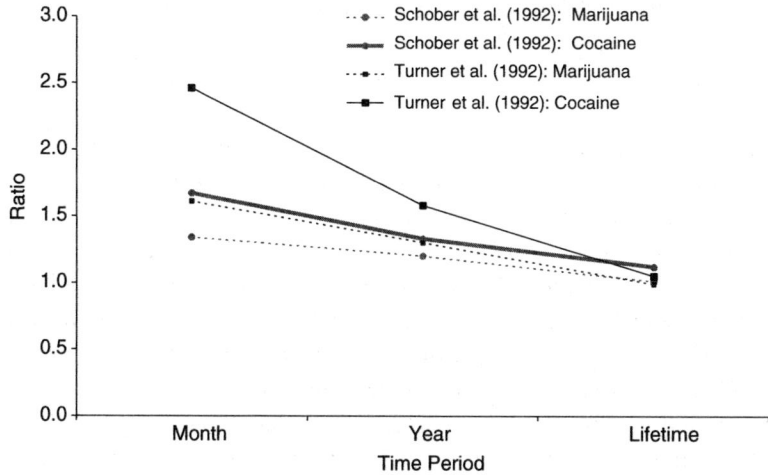

FIGURE 2-1 Drug reporting.

administration, Figure 2-1 plots the ratio between the level of illicit drug use reported when survey questions are self-administered to the level reported when interviewers administer the questions. For example, if 6 percent of respondents report using cocaine during the previous year under self-administration but only 4 percent report using cocaine under interviewer administration, the ratio would be 1:5. The data are from two of the largest mode comparisons done to date: one carried out by Turner, Lessler, and Devore (1992) and the other by Schober, Caces, Pergamit, and Branden (1992). The ratios range from a little over 1:0 to as high as 2:5— which is to say that self-administration more than doubled the reported rate of illicit drug use.

Tourangeau, Rips, and Rasinski (2000:Table 10.2) reviewed a number of similar mode comparisons; they found a median increase of 30 percent in the reported prevalence of marijuana use with self-administration and similar gains for cocaine use. They also summarize the evidence that self-administration improves reporting about other sensitive topics, including sexual partners, abortion, smoking, and church attendance.

Mode, Privacy, and the Presence of Third Parties

It is natural to think that at least some of the gains from self-administration result from the reduced risk of disclosure to other household members, but there is surprisingly little evidence that the presence of other household members has much effect on what respondents report during interviews. Interviews are often conducted under less than ideal conditions, and, although most survey organizations train their interviewers to try to find private settings for the interviews, other household members are often present. For example, Silver and colleagues examined the proportion of interviews done for the American National Election Studies (ANES) in which other household members were present. The ANES is a series of surveys funded by the National Science Foundation and carried out by the University of Michigan's Survey Research Center. The proportion varied somewhat from one survey to the next, but roughly half of all interviews conducted between 1966 and 1982 were done in the presence of another household member (Silver et al., 1986). Similarly, Martin and colleagues (1986) noted that some 58 percent of NCS interviews were conducted within earshot of someone other than the interviewer and respondent (for a more recent estimate, see Coker and Stasny, 1995).

Silver and colleagues looked at whether the presence of other people during interviews affected the overreporting of voting. In many jurisdictions, whether someone voted is a matter of public record, so it is a relatively easy matter to determine the accuracy of reports about voting. Voting is a socially desirable behavior, and many nonvoters—roughly a quarter, according to Silver and company—nonetheless report that they voted during the most recent election. What is somewhat surprising is that the rate of overreporting did not vary as a function of the privacy of the interview. A number of other national surveys have also recorded whether other people are present during the interviews, but researchers who have examined these data have found little evidence that the presence of others affects reports about such potentially sensitive topics as sexual behavior (Laumann et al., 1994) or illicit drug use (Schober et al., 1992; Turner et al., 1992). Smith's (1997) review concludes that in general the effects of the presence of third parties during interviews are minimal.

There are several possible explanations for the absence of third-party effects. As Martin and colleagues note, other household members may remember relevant incidents that a respondent has forgotten, offsetting any inhibiting effect their presence has. When another household member already knows the sensitive information, his or her presence may make it more difficult for the respondent to withhold it from the interviewer. In addition, interviewers are probably more likely to do interviews with other household members who are around when the respondent seems unconcerned about their presence.

Few of the studies examining the impact of third parties have used experimental designs that systematically varied the privacy of the interview. Still, in crime surveys one would expect the presence of family members to have an impact, particularly on reports involving domestic violence; similarly, in surveys on rape the presence of family members is likely to inhibit reports of spousal rape, if not rape in general. In fact, there is some recent evidence that the presence of a spouse during an interview is associated with reduced reporting of rape and domestic violence (Coker and Stasny, 1995). Since more than half of the NCVS interviews are conducted with at least one other person present, it is likely that estimates of certain victimizations are affected by the presence of a third party.

Variations Across Methods of Self-Administration

The lack of evidence of third-party effects—in contrast to the large and consistent effects of self-administration—suggests that respondents are less concerned about the reactions of other household members than about those of the interviewer. Tourangeau, Rips, and Rasinski (2000) argue that what survey respondents worry about is that they will be embarrassed during an interview; the prototypical embarrassment situation involves disclosure to strangers rather than friends or family members (Tangney et al., 1996). Despite their efforts to create rapport with respondents, when interviewers ask the questions and record the answers, it raises the specter they will react negatively to what the respondent tells them and embarrass the respondent. That risk is reduced when respondents can answer on paper or interact directly with computers.

It is possible that some of the newer computer-assisted forms of self-administration confer an even greater sense of privacy than traditional paper-and-pencil self-administered questionnaires (SAQs). Turner and colleagues compared data collected via ACASI with data from a traditional SAQ; their experiment involved random subgroups of participants from the National Survey of Adolescent Males (Turner et al., 1998). On some of the most sensitive questions (such as one asking about receptive oral sex), ACASI yielded four times as many reports as the paper SAQ (see Figure 2-2). Tourangeau and Smith (1996) also found increased reporting for some items under ACASI relative to CASI.

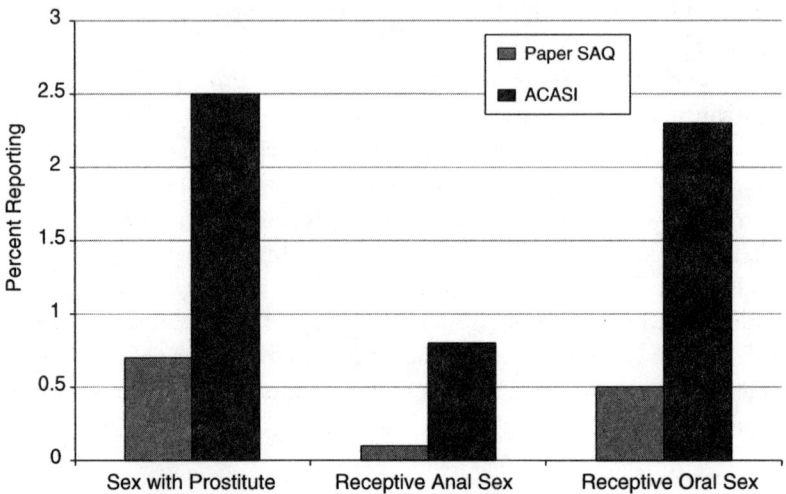
FIGURE 2-2 Reporting by method of self-administration.

What all of these findings suggest is that crime reports—especially reports about victimizations involving crimes that carry stigma—could be dramatically affected by the mode of data collection. Self-administration of the questions is likely to increase the number of rape victimizations reported; on the other hand, it may sharply reduce reports of defensive gun use, since defensive gun use is likely to be seen as a positive or socially desirable response to crime (as Hemenway, 1997, argues).

IMPACT OF CONTEXT

The context of a survey question encompasses a wide range of potential influences, including the perceived climate of opinion as the survey is being done, the auspices under which the survey is conducted, the purpose of the survey (as it is presented to the respondent), the topics preceding a given question, characteristics of the interviewers related to the survey topic, the physical setting for the interview, and even the weather (see Schwarz and Clore, 1983). Most of the research on context effects has focused more narrowly on the impact of prior items on answers to later questions, but even under this restricted definition of survey context, context has a range of effects on subsequent items, altering the overall direction of answers to specific questions or changing the relationship among survey items.

Setting of the Interview

Partly because of concerns about the inhibiting effects of the presence of third parties during interviews, some survey researchers have experimented with changing the overall context by conducting survey interviews in settings outside respondents' homes. For example, Jobe and colleagues did an experiment in which half of the sample of women in their study were interviewed at a neutral site and the other half were interviewed at home (Jobe et al., 1997). The interview touched on a number of sensitive topics, including sexual partners, sexually transmitted diseases, illicit drug use, and abortion. No consistent impact on reporting by interview site was found. A second study by Mosher and Duffer (1994; see also Lessler and O'Reilly, 1997) found an increase in the number of abortions reported when interviews were conducted outside the home; unfortunately, the outside-the-home group showed no significant increase over an in-home group given the same incentive to take part in the study.

Still, a couple of more recent studies suggest that answers to survey questions can be affected by the physical setting in which the data are collected. In one study (Beebe et al., 1998), students were more likely to admit illicit drug use, fighting, and other sensitive behaviors on a paper SAQ than on a computer-administered version of the same questions. Both versions were administered in a school setting. As noted earlier, computerized self-administration generally preserves or increases the gains from paper self-administration, but Beebe and colleagues argue that various features of the survey setting can reduce the sense of privacy that computers usually confer and, as a result, affect the answers. In their study the computerized questions were administered in the school's computer lab on networked computers via terminals that were located next to one another. Any of these features may have reduced the apparent privacy of the data collection process.

In another study (Moon, 1998), respondents were sensitive to the description of the computer that administered the questions. When the respondents thought a distant computer was administering the questions to them over a network, they gave more socially desirable answers than when they thought the computer directly in front of them was administering the questions. Even when respondents do not have to interact with an interviewer, the characteristics of the data collection setting can enhance or reduce the sense of privacy and affect the answers that respondents give.

Context and the Inferred Purpose of the Question

Survey items, like everyday utterances, may convey a lot more to the respondent than they literally state. Consider this item from the General Social Survey (GSS), a national attitude survey carried out by the National Opinion Research Center since 1972: "Are there any situations you can imagine in which you would approve of a policeman striking an adult male citizen?" Taken literally, the item invites a positive answer. After all, anyone who has ever seen a cop show on television would have little trouble conjuring up situations in which a police officer would be fully justified in using force—say, in subduing an escaping criminal. Yet in recent years, fewer than 10 percent of respondents have answered "yes" to this GSS question. Clearly, respondents—a national cross-section of adults in the United States—do not take the question literally but base their answers on their reading of the intent behind the question. They seem to infer that the item is intended to tap their feelings about police brutality or police violence, not to test their imaginative powers. In the same way, in everyday life, what is literally a question may in fact convey a request ("Can you pass the salt?") or a command ("Will you close the door on your way out?"). Part of the task of the survey respondent is to determine the intent of the question, to ferret out the information it is really seeking.

Interviews as Conversations

Unfortunately, in the process of interpreting survey questions, respondents often draw unintended inferences about their meaning. They draw these incorrect inferences because they carry over into the survey interview interpretive habits developed over a lifetime in other settings, such as conversations (Schaeffer, 1991). In a conversation, usage is often loose—words and phrases may be used in a nonstandard way or intended nonliterally—but little harm is done because the participants can interact freely to make sure they understand each other. Speakers pause to be sure their listeners are still with them; listeners use phrases like "okay" and "uh-huh" to signal that they are following (Clark and Schaefer, 1989; Schober, 1999). The resources available in conversation to forestall or repair misunderstandings are sharply curtailed in survey interviews, which are typically standardized to eliminate much of this unscripted interaction (Suchman and Jordan, 1990).

Based on their experiences in other settings, respondents may have incorrect expectations about survey interviews. In a conversation a question such as, "Has anyone attacked or threatened you in the last year?" might be intended simply to determine the general topic for the next few exchanges. It would be quite reasonable for the listener to respond, "Well, a few years ago someone threw a beer bottle at my car." The incident may not really constitute an attack and it falls outside the time frame set by the question, but it is in the ballpark, which is all that is required for everyday conversation. In a survey, though, such an item is generally intended as a request for exact and accurate information; "last year" is typically intended to refer to the date exactly one year ago. Many survey respondents do not seem to realize that surveys are actually seeking this sort of precise information (Cannell et al., 1968).

Grice's Conversational Maxims

Conversations are partly guided by what the philosopher Paul Grice called the *cooperative principle*—the unstated assumption that the participants in a conversation are trying to make their contributions useful. Grice (1989) distinguished four specific manifestations of the cooperative principle—four conversational maxims that shape everyday conversations and serve as a kind of etiquette for them:

- the *maxim of quantity*, which requires us to make our contribution as informative as necessary, but not more informative;
- the *maxim of quality*, which says to tell the truth and to avoid statements we cannot support;
- the *maxim of relation*, which enjoins us to stick to the topic; and
- the *maxim of manner*, which tells us to be clear and to avoid obscurity.

When a speaker seems to violate one of these norms, the listener generally cuts him or her some interpretive slack, drawing the inferences that are needed to continue to see the speaker as cooperative. For example, if the speaker seems to have changed the topic, the listener will assume there is some connection with what came before and will attempt to identify the thread.

Surveys violate the Gricean maxims all the time. They flit from one subject to another without any of the transitional devices that would mark a change of subject in ordinary conversation. Sometimes respondents make

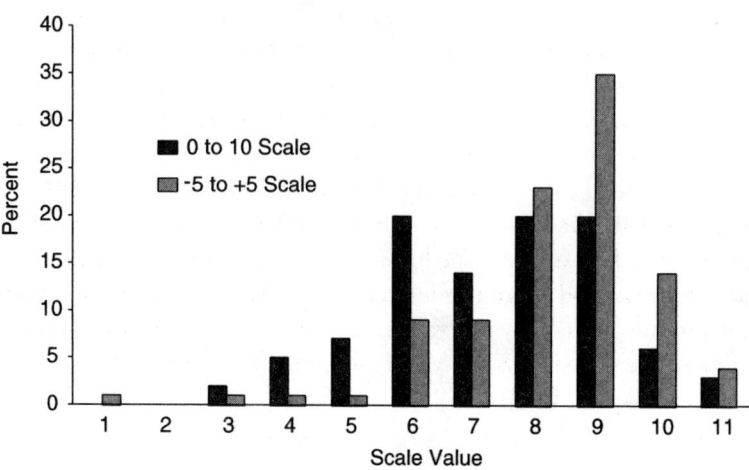

FIGURE 2-3 Impact of category labels.

the adjustment and behave as though Grice's maxims do not apply to surveys, but researchers have marshaled a large of number of examples in which survey respondents apparently apply the Gricean maxims anyway, drawing unintended inferences from incidental features of the survey questions or their order. A study by Schwarz and colleagues illustrates the problem (Schwarz et al., 1991). The study compared two items asking respondents how successful they had been in life; the only difference between the items was in the labeling of the scale points. One group got an 11-point scale in which the scale values ranged from 0 to 10; the other group got scale values that ranged from –5 to +5. These values were displayed to the respondents in a show card. Figure 2-3 presents the key results—a sharp change in the distribution of the answers depending on the numerical labels on the scale points. Respondents who got the scale labeled with negative values shied away from that end of the scale. If "0" conveys essentially no success in life, "–5" suggests catastrophic failure.

Respondents may also draw unintended inferences based on other features of the questions or their order. When the questionnaire juxtaposes two similar items ("How much did you like the food on your most recent visit to McDonald's?" "Overall, how satisfied were you with your meal at McDonald's?"), respondents may think they are intended to convey quite different questions (Strack et al., 1991). After all, according to the maxim of quantity, each contribution to an ongoing conversation is supposed to convey *new* information. As a result, the correlation between the two items can be sharply reduced as compared to when they are separated in the interview (see also Schwarz et al., 1991; Tourangeau et al., 1991). Surveys often use closed questions that offer respondents a fixed set of answer categories. Respondents may infer that the options offered by the researchers reflect their knowledge about the distribution of the answers; the middle option, they may reason, must represent the typical answer (Schwarz et al., 1985). Tourangeau and Smith (1996) found that the ranges used in the answer categories even affected respondents' reports about the number of sexual partners they had in the past five years.

Cooperativeness, Satisficing, and Survey Auspices

Respondents interpret the questions in surveys not necessarily in the way the surveys' designers had hoped. They sometimes lean on relatively subtle cues—the numbers assigned to the scale points, the order of the questions, the arrangement of the response categories—to make inferences about the intent behind the questions and about their job as respondents. Interpretive habits acquired over a lifetime are applied—not always appropriately—to the task of understanding and answering survey questions. A key point left out of the discussions of the defensive gun use and rape prevalence controversies is how the presentation of a survey to the respondents may have affected their understanding of the survey task and their interpretation of specific questions.

Over the past 15 years or so, survey researchers have begun to apply findings from the cognitive sciences, particularly cognitive psychology, in a systematic program to understand reporting errors in surveys (see Sirken et al., 1999; Sudman et al., 1996; Tourangeau et al., 2000, for recent reviews of this work). One theme that clearly emerges from this literature is that respondents take whatever shortcuts they can to reduce the cognitive effort needed to answer the questions (see, for example, Krosnick and Alwin, 1987; Krosnick, 1991). There are probably several reasons for this.

In the first place, as Cannell, Fowler, and Marquis (1968) noted more than 30 years ago, respondents may simply not realize they are supposed to work hard and provide exact, accurate information. One of the reasons that bounding procedures, like the one used in the NCVS, are thought to be effective is that they convey to respondents the need for precision (cf. Biderman and Cantor, 1984). Some of the gains from bounding can be realized simply by using an exact date in the question (e.g., "Since June 4 . . ." as opposed to "During the past month . . ."; see Loftus and Marburger, 1983) or by dividing the recall period into shorter periods and asking about each one separately (Sudman et al., 1984). Each of these methods of defining the recall period implicitly conveys the need for precision. In everyday conversation, of course, participants typically adopt a much looser criterion in framing answers to questions.

Even if respondents do realize they are supposed to be exact and accurate, they may be unwilling or unable to oblige. Consider the questions used by Kleck and colleagues, cited earlier, to measure defensive gun use. The key question asks about a rather long time period—five years. It covers both the respondents themselves and other members of their households; it lists several possible relevant scenarios ("for self-protection or for the protection of property at home, work, or elsewhere"); and it notes several exclusions (incidents involving "military service and police security work"). The demands of interpreting this item, mentally implementing its complicated logical requirements, and searching memory for relevant incidents over such a long period are likely to exceed the working memory capacity of many respondents, even well-motivated ones (Just and Carpenter, 1992).

It is not hard to believe that many respondents do not take the question literally. After all, taking the question literally might entail putting the telephone down and canvassing other household members; just reconstructing what you were doing five years ago might require more thought and a longer pause than is usually considered appropriate for a telephone conversation. Chances are that many respondents opt out of the literal requirements of the question and adopt a less exacting, "satisficing" criterion; they try to formulate an answer that is good enough to satisfy the spirit of the question rather the letter. Of course, a less than perfect answer will not always yield a false positive response. Some respondents may be unaware of or overlook gun use by other household members. But as we argued earlier, even random errors are likely to yield an upward bias in this case because, according to all the estimates, gun use is still very rare.

In determining the real intent behind survey questions, respondents may rely on a variety of cues, but the apparent purpose and auspices of a survey are likely to be among the most important. What are respondents likely to conclude about the intent of a survey like the NCVS? The study is conducted by the Bureau of the Census, the source of many important official statistics, on behalf of another federal agency, the Bureau of Justice

Statistics. The interviewers wear identification badges and carry an advance letter printed on government stationery that explains the purpose of the survey. All of these cues—even the very name of the survey—are likely to convey to the respondents that the topic is *crimes*, narrowly construed. Even when the questionnaire probes for incidents that might not ordinarily be seen as crimes—say, domestic violence or unwanted sex when the victim was drunk—respondents may not take the items literally, inferring a narrower scope to the questions than is actually intended. Respondents want to cooperate with the perceived demands of the interview, but they do not want to work hard at it. When construing the topic narrowly fits their impression of the intent of the study and allows them to get through the interview without working very hard, that is what most respondents are likely to do.

If the easy way to meet the apparent demands of the NCVS is to construe the questions narrowly, omitting borderline incidents, atypical victimizations, and incidents that may fall outside the time frame for the survey, respondents may adopt the opposite approach in many other surveys. For example, many of the rape surveys cited by Koss and by Fisher and Cullen (2000) are local in scope, involving a single college or community (see, e.g., Table 1 in Koss, 1993); they generally do not have federal sponsorship and are likely to appear rather informal to the respondents, at least as compared to the NCVS. Many of the surveys are not bounded and cover very long time periods (e.g., the respondent's entire life). The names of these surveys (e.g., Russell, 1982, called her study the Sexual Victimization Survey; Koss's questionnaire is called the Sexual Experiences Survey), their sponsorship, their informal trappings, their content (numerous items on sexual assault and abuse), and their long time frame are likely to induce quite a different mindset among respondents than that induced by the NCVS.

Many of the rape studies seem to invite a positive response; indeed, their designs seem predicated on the assumption that rape is generally underreported. It seems likely that many respondents in these surveys infer that the intent is to broadly document female victimizations, even though the items used are very explicit. The surveys and the respondents both seem to cast a wide net. When Fisher and Cullen (2000) compared detailed reports about incidents with responses to the rape screening items in the National Violence Against College Women Study, they classified only about a quarter of the incidents mentioned in response to the rape screening items as actually involving rapes. (Additional incidents that qualified as rapes were reported in response to some of the other screening items as well.) Respondents want to help; they have volunteered to take part in the survey and are probably generally sympathetic to the aims of the survey sponsors. When being helpful seems to require reporting relevant incidents, they report whatever events seem most relevant, even if they do not quite meet the literal demands of the question. When the surveys do not include detailed follow-up items, there is no way to weed out reports that meet the perceived intent but not the literal meaning of the questions.

Of course, we do not *know* how the sponsorship and other trappings of these surveys affect reporting, but it seems to be well worth exploring. Other aspects of survey context are known to have a large impact on reporting; it would be easy to find out how variations in the way a survey is presented to respondents affect their perceptions of the task and the answers they ultimately provide.

Impact of Prior Questions

The presentation of a survey—in the survey introduction, the advance letter, even the name of the study—can shape respondents' understanding of their task and their interpretation of individual questions in the interview. Prior questions in the survey can have a similar impact, affecting what respondents think about as they answer later questions and how they understand the questions (see Tourangeau, 1999, for a review). The impact of prior items on answers to later ones reflects several distinct mechanisms, two of which are especially relevant here: The earlier items sometimes provide an interpretive context for later questions, and sometimes they trigger the recall of information useful in answering later items.

According to Grice's maxim of relation, the participants in a conversation are supposed to stick to the topic; they are not supposed to shift gears without giving proper warning. For the most part, survey questions follow these rules, signaling shifts in topics with introductions or transitional phrases ("The next few items are about . . ."). Relying on the maxim of relation, survey respondents may look to prior items to clarify the meaning of a new question. For example, a study by Strack and his colleagues (1991) asked German students if they would support an "educational contribution." German universities are tuition free, and it was not clear from the question whether the educational contribution was to be given to the students in the form of financial aid or taken from them in the form of tuition. When the item came after a question on college tuition in the

United States, support for the "educational contribution" dropped; when it followed a question on government aid to students in Sweden, support increased. Respondents apparently looked to the previous question to decide what "educational contribution" meant. In this case the maxim of relation led respondents to see more similarity than they should have between adjacent items.

Alternatively, the maxim of quantity ("Be informative") may lead them to make the opposite error—to exaggerate the differences among successive questions. For instance, in one study, respondents were asked to evaluate the economy in their state and the economy in their local communities (Mason et al., 1995). They tended to cite different reasons in explaining their evaluations of the local economy when that item came first than when it followed the question on the state economy. Apparently, they were reluctant to give the same rationale ("good job prospects" or "growth in local industries") to explain both answers and so were forced to come up with new considerations to justify their views about the local economy.

Another way that prior items can affect answers to later questions is by reminding respondents of things they would not have otherwise recalled. The process of retrieving information from memory is partly deliberate and partly automatic. The deliberate part consists of generating incomplete descriptions of the incident in question. In the case of survey items, we might start with the description given in the question itself ("Let's see, a time when I was attacked or threatened") and supplement it with inferences or guesses about the sought-after incidents ("Didn't something happen on Gough Street a couple of months ago?"). The automatic component consists of our unconsciously tracing links between associated ideas in memory; thinking about one incident makes it easier for us to retrieve other events that are associated with it.

In the jargon of cognitive psychology, "activation" spreads automatically from a single concept to related concepts (e.g., Anderson, 1983). This component is automatic in the sense that it operates without our willing it, indeed without our being aware of it. Because of the spread of activation, prior questions can set retrieval processes in motion that alter our answers to later questions. One example of this kind of context effect was apparently found during one of the developmental studies for the NCS. Respondents who had first answered a series of questions designed to assess their fear of crime reported more victimizations than their counterparts who answered the victimization items first (Murphy, 1976). The fear of crime items apparently facilitated recall of victimizations.

These effects of prior items on the interpretation and retrieval process for subsequent questions means that asking the "same" question in two different questionnaires will not necessarily yield the same answers. As Koss (1993) argues, the focus of the NCVS on *criminal* victimizations (along with other cues in that study) may promote a narrow interpretation of the type of incidents to be reported; in addition, the screening items in the NCVS may serve as relatively poor retrieval cues for incidents the respondents do not necessarily think of as crimes. On the other hand, some of the items used in the Sexual Experiences Survey (Koss et al., 1987) and later rape surveys—"Have you given in to sex play (fondling, kissing, or petting, but not intercourse) when you didn't want to because you were overwhelmed by a man's continual arguments and pressure?"—may help prompt fuller recall of more serious incidents but may also suggest that almost any unwanted sexual experience is of interest, encouraging overreporting.

Bounding

Another procedure used in some crime surveys may also help frame later questions for respondents and trigger the recall of relevant events; this is the review of incidents reported in the previous round as part of the bounding procedure. The purpose of bounding is to prevent respondents from reporting incidents that actually occurred before the start of the recall period; this type of error is known as telescoping in the survey literature. Although more sophisticated theories of telescoping have been proposed, it mostly appears to reflect our poor memory for dates (see, for example, Baddeley et al., 1978; Thompson et al., 1996). Because telescoping errors are common, bounding can have a big effect on the level of reporting in a survey. In fact, Neter and Waksberg's (1964) initial demonstration of the benefits of bounding indicated that about 40 percent of all household repairs and more than half of related expenditures were reported in error in unbounded interviews. Table 2-1 summarizes the results of the Neter and Waksberg (1964) study, along with a series of studies by Loftus and Marburger (1983) that explored alternative procedures for bounding the recall period.

Loftus and Marburger used several procedures to define the beginning of the recall period. They carried out two of their experiments exactly six months after the eruption of Mt. Saint Helens; in those studies, they compared answers to questions that began "Since the eruption of Mt. St. Helens

..." with parallel items that began "During the last six months. . . ." The eruption of Mt. Saint Helens served as what Loftus and Marburger called a temporal landmark. In subsequent experiments, they used New Year's Day to mark off the boundary of the recall period or asked respondents to generate their own personally significant landmark event near the beginning of the recall period. As Table 2-1 indicates, whether bounding takes the form of reviewing with the respondents what they already reported in the previous interview (as in Neter and Waksberg, 1964), providing them with a public landmark event, like the eruption of Mt. St. Helens or New Year's Day (Loftus and Marburger, 1983), or asking them to generate their own personal landmark events (Loftus and Marburger, 1983:Experiment 3), bounding sharply reduces the level of reporting.

Bounding probably has several useful effects. First, as Biderman and Cantor (1984) noted, it helps communicate the importance of precision; just mentioning the specific date that marked the beginning of the recall period had a noticeable impact on the number of victimizations reported in Loftus and Marburger's final experiment. It is quite likely that many respondents begin survey interviews thinking that it will be cooperative for them to mention incidents related to the topic of the interview, even if those incidents do not meet all the requirements set forth in the questions. Bounding procedures help alter this expectation. But the impact of bounding seems to go beyond its role in socializing the respondent to the task's requirements. A variety of evidence suggests that people are much better at reconstructing the relative order of different events than recalling their exact dates (e.g., Friedman, 1993). Bounding converts the temporal judgment respondents have to make from an absolute one (Does the event fall in this period?) to a relative one (Did the event occur before or after the bounding event?); relative judgment is a lot more accurate.

Bounding procedures can serve still another function—both previously reported incidents and landmark events can serve as powerful retrieval cues. When people are asked to remember events from a given period, they tend to recall incidents that fall near temporal boundaries, such as the beginning of the school year or major holidays (Kurbat et al., 1998; Robinson, 1986). Major temporal periods are an important organizing principle for our autobiographical memories; if our memories were written autobiographies, they would be made up of chapters corresponding to each of the major periods of our lives. The boundaries that separate these periods are powerful retrieval cues. Similarly, the review of events reported in an earlier interview

TABLE 2-1 Impact of Bounding Procedures

Study	Bounding Procedure	Ratio of Events Reported: Unbounded over Bounded	
Neter and Waksberg (1964)	Prior interview	Expenditures	1.40
		Jobs	1.55
Loftus and Marburger (1983)			
Experiment 1	Landmark event	Any victimizations	6.15
Experiment 2	Landmark event	Victim of theft	1.51
		Victim of assault	1.52
		Reported crime	1.22
Experiment 3	Personal landmark	Any victimizations	5.50
Experiment 4	New Year's Day	Any victimizations	2.00
Experiment 5	New Year's Day	Any victimizations	2.52
	Specific date	Any victimizations	1.32

may trigger the recall of similar or related incidents since then. Bounding procedures improve the accuracy of recall, helping respondents weed out ineligible events and remember eligible ones. As Table 2-1 suggests, the net effect can be dramatic.

CONCLUSIONS: SOURCES OF DIFFERENCES ACROSS SURVEYS

Most papers that examine discrepancies across surveys are limited to speculating about the sources of the differences in the results, and this paper is no exception. (McDowall et al., 2000, and Fisher and Cullen, 2000, *are* exceptions—they present evidence testing specific hypotheses about why different procedures gave different results.) Throughout, we have offered conjectures about the variables that affect reporting in crime surveys. In this final section we try to be a little more explicit about the variables we think are the key ones. One theme that runs through our discussion is that both overreporting and underreporting are possible; it simply cannot be taken for granted that relatively rare events, like defensive gun use—or even very sensitive ones, like rape victimizations—will necessarily be underreported in surveys. Respondents can only make the errors it is logically possible for them to make; if most of them have not in fact experienced the target events, they can only overreport them. Moreover, as the work of Loftus reminds us, forgetting does not necessarily make us underreport events. Forgetting when something happened or what exactly took place can lead us to report events that do not really count. And the same cues that can help us remember an event can also encourage us to report incidents that do not meet the requirements of a survey's questions.

A variable that has been relatively neglected in discussions of crime reporting has been the *mode* of data collection. There is strong evidence that self-administration produces fuller reporting of sensitive behaviors, sometimes dramatically so (as in Figure 2-1). Several new methods of computerized self-administration have become available over the past 10 years; these new methods have greatly extended the range of situations in which self-administration can be used and in some cases have sharply increased levels of reporting (e.g., see Figure 2-2). The new technologies can be used in conjunction with face-to-face (Tourangeau and Smith, 1996; Turner et al., 1998) or telephone interviews (Phipps and Tupek, 1990; Turner et al., 1998), or they can be used to administer stand-alone surveys via e-mail (e.g., Kiesler and Sproull, 1986) or the Internet (Dillman et al., 1998).

Our first hypothesis, then, is that self-administration will dramatically increase reports of some types of crime, particularly those that carry stigma and those perpetrated by other household members; self-administration will reduce reports of incidents that put the respondent in a favorable light, including perhaps defensive gun use. A related hypothesis involves the presence of other household members; for the topics raised in crime surveys, we believe that the presence of other household members must make a difference (at least for crimes involving domestic violence). Crime surveys may prove to be a notable exception to the rule that the presence of third parties during an interview does not have much effect on a respondent's answers (see Coker and Stasny, 1995, for some evidence supporting this conjecture).

We also offer several hypotheses about the effects of the context of a survey, construing context broadly to include not only the previous items in the questionnaire but also the packaging of the survey to the respondent and the procedures used to bound the recall period. Our first hypothesis is that the apparent topic of the survey, the survey's sponsorship, the organization responsible for collecting the data, the letterhead used on advance letters, and similar procedural details will affect respondents' views about the need for accuracy in reporting and about the type of incidents they are supposed to report. It is easy to imagine an experiment that administers the same questions to all respondents but varies the framing of the survey. Such an experiment would examine how reports were affected by the framing of the survey; in addition, it might also compare respondents' judgments as to whether they were *supposed* to report hypothetical incidents in vignettes describing borderline cases. Our guess is that the packaging of the survey will have a big impact on how respondents classify the incidents depicted in the vignettes.

Our next hypothesis is that the context provided by earlier questions will have effects similar to those of the context provided by the external trappings of the survey. A rape survey loaded with crime items is likely to lead respondents to omit sexual victimizations that do not seem crimelike; a survey loaded with items on sexual victimizations will lead respondents to report incidents that are not, strictly speaking, rapes. Respondents want to help out by providing relevant information, but they are accustomed to the looser standards of conversation and take cognitive shortcuts to reduce the demands of the questions. As a result, it is important to gather detailed information about each incident that respondents report. Even when elaborate and explicit screening items are used, the researchers' classification of an incident will not necessarily agree with the respondent's (Fisher and Cullen, 2000).

We discussed one additional contextual variable—the bounding procedure used to frame the recall period for the survey items. Our final hypothesis is that the exact bounding procedure a survey uses will sharply affect the final estimates. Surveys that ask about events during a vaguely defined recall period (e.g., the Kleck surveys on defensive gun use) will yield more reports than surveys that take the trouble to bound the recall period more sharply. An exact date is good and a landmark event is better. By itself a prior interview may not be all that effective as a bounding event; our final hypothesis is that a full bounding procedure that includes a review of the incidents reported in the previous interview will reduce reporting relative to the more truncated procedure used in many surveys that simply instructs respondents to report incidents that occurred since the last interview. Compared to temporal or personal landmarks, the prior interview may not mark off the relevant time period very clearly.

Of course, lots of things affect reporting in surveys. Crime surveys are at an added disadvantage because many of their questions involve particularly stigmatized or traumatic events, such as rape, that respondents may simply not want to discuss. This is why it is especially important to do as much as possible to uncover the effects of those factors that are within the control of the researchers. We have tried to focus on a few of the variables—the mode of interviewing, the setting of the interview, the framing of the survey, and the context of the key items—that we think may have a big impact on reporting in crime surveys. These are variables that have been shown to have effects large enough to account for the very large differences in results across different surveys. Unfortunately, we will not know whether these are the culprits until someone does the right experiments.

REFERENCES

Anderson, J.R.
  1983 *The Architecture of Cognition.* Cambridge, MA: Harvard University Press.
Baddeley, A.D., V. Lewis, and I. Nimmo-Smith
  1978 When did you last . . .? In *Practical Aspects of Memory,* M.M. Gruneberg, P.E. Morris, and R.N. Sykes, eds. London: London Academic Press.
Beebe, T.J., P.A. Harrison, J.A. McRae, R.E. Anderson, and J.A. Fulkerson
  1998 An evaluation of computer-assisted self-interviews in a school setting. *Public Opinion Quarterly* 11:623-632.
Biderman, A.
  1980 *Report of a Workshop on Applying Cognitive Psychology to Recall Problems of the National Crime Survey.* Washington, DC: Bureau of Social Science Research.
Biderman, A., and D. Cantor
  1984 A longitudinal analysis of bounding, respondent conditioning, and mobility as sources of panel bias in the National Crime Survey. In *Proceedings of the American Statistical Association, Survey Research Methods Section.* Washington, DC: American Statistical Association.
Bradburn, N.M., and S. Sudman
  1979 *Improving Interview Method and Questionnaire Design.* San Francisco: Jossey-Bass.
Burton, S., and E. Blair
  1991 Task conditions, response formulation processes, and response accuracy for behavioral frequency questions in surveys. *Public Opinion Quarterly* 55:50-79.
Cannell, C., F.J. Fowler, and K. Marquis
  1968 The influence of interviewer and respondent psychological and behavioral variables on the reporting in household interviews. *Vital and Health Statistics* 2:26.
Cannell, C., P. Miller, and L. Oksenberg
  1981 Research on interviewing techniques. Pp. 389-437 in *Sociological Methodology,* S. Leinhardt ed. San Francisco: Jossey-Bass.

Clark, H.H., and E.F. Schaefer
  1989   Contributing to discourse. *Cognitive Science* 13:259-294.
Coker, A.L., and E.A. Stasny
  1995   *Adjusting the National Crime Victimization Survey's Estimates of Rape and Domestic Violence for "Gag" Factors.* Washington, DC: U.S. Department of Justice, National Institute of Justice.
Dillman, D.A., R.D. Tortora, J. Conradt, and D. Bowker
  1998   Influence of plain vs. fancy design on response rates for Web surveys. In *Proceedings of the Survey Research Methods Section, American Statistical Association.* Washington, DC: American Statistical Association.
Fisher, B.S., and F.T. Cullen
  2000   *Measuring the Sexual Victimization of Women: Evolution, Current Controversies, and Future Research.*
Friedman, W.J.
  1993   Memory for the time of past events. *Psychological Bulletin* 113:44-66.
Grice, H.P.
  1989   *Studies in the Way of Words.* Cambridge, MA: Harvard University Press.
Hemenway, D.
  1997   The myth of millions of annual self-defense gun uses: A case study of survey overestimates of rare events. *Chance* 10:6-10.
Jobe, J.B., W.F. Pratt, R. Tourangeau, A. Baldwin, and K. Rasinski
  1997   Effects of interview mode on sensitive questions in a fertility survey. Pp. 311-329 in *Survey Measurement and Process Quality,* L. Lyberg, P. Biemer, M. Collins, E. de Leeuw, C. Dippo, N. Schwarz, and D. Trewin, eds. New York: Wiley.
Just, M.A., and P.A. Carpenter
  1992   A capacity theory of comprehension. *Psychological Review* 99:122-149.
Kiesler, S., and L. Sproull
  1986   Response effects in the electronic survey. *Public Opinion Quarterly* 50:402-413.
Kleck, G.
  1991   *Point Blank: Guns and Violence in America.* Hawthorne, NY: Aldine de Gruyter.
Kleck, G., and M. Gertz
  1995   Armed resistance to crime: The prevalence and nature of self-defense with a gun. *Journal of Criminal Law and Criminology* 86:150-187.
Koss, M.
  1992   The underdetection of rape. *Journal of Social Issues* 48:63-75.
  1993   Detecting the scope of rape: A review of prevalence research methods. *Journal of Interpersonal Violence* 8:198-222.
  1996   The measurement of rape victimization in crime surveys. *Criminal Justice and Behavior* 23:55-69.
Koss, M., C.A. Gidycz, and N. Wisniewski
  1987   The scope of rape: Incidence and prevalence of sexual aggression and victimization in a national sample of higher education students. *Journal of Consulting and Clinical Psychology* 55:162-170.
Krosnick, J.A.
  1991   Response strategies for coping with the cognitive demands of attitude measures in surveys. *Applied Cognitive Psychology* 5:213-236.

Krosnick, J.A., and D. Alwin
  1987 An evaluation of a cognitive theory of response-order effects in survey measurement. *Public Opinion Quarterly* 51:201-219.
Kurbat, M.A., S.K. Shevell, and L.J. Rips
  1998 A year's memories: The calendar effect in autobiographical recall. *Memory and Cognition* 26:532-552.
Laumann, E., J. Gagnon, R. Michael, and S. Michaels
  1994 *The Social Organization of Sexuality: Sexual Practices in the United States.* Chicago: University of Chicago Press.
Lehnen, R.G., and W. Skogan
  1981 *The National Crime Survey: Working Papers. Volume 1: Current and Historical Perspectives.* Washington, DC: Bureau of Justice Statistics.
Lessler, J.T., and J.M. O'Reilly
  1997 Mode of interview and reporting of sensitive issues: Design and implementation of audio computer-assisted self-interviewing. Pp. 366-382 in *The Validity of Self-Reported Drug Use: Improving the Accuracy of Survey Estimates*, L. Harrison and A. Hughes, eds. Rockville, MD: National Institute on Drug Abuse.
Loftus, E.F., and W. Marburger
  1983 Since the eruption of Mt. St. Helens, has anyone beaten you up? Improving the accuracy of retrospective reports with landmark events. *Memory and Cognition* 11: 114-120.
Lynch, J.P.
  1996 Clarifying divergent estimates of rape from two national surveys. *Public Opinion Quarterly* 60:410-430.
Martin, E., R.M. Groves, J. Matlin, and C. Miller
  1986 *Report on the Development of Alternative Screening Procedures for the National Crime Survey.* Washington, DC: Bureau of Social Science Research.
Mason, R., J. Carlson, and R. Tourangeau
  1995 Contrast effects and subtraction in part-whole questions. *Public Opinion Quarterly* 58:569-578.
McDowall, D., and B. Wiersema
  1994 The incidence of defensive firearm use by U.S. crime victims: 1987 through 1990. *American Journal of Public Health* 84:1982-1984.
McDowall, D., C. Loftin, and S. Presser
  2000 Measuring civilian defensive firearm use: A methodological experiment. *Journal of Quantitative Criminology* 16:1-19.
Means, B., G.E. Swan, J.B. Jobe, and J.L. Esposito
  1994 An alternative approach to obtaining personal history data. Pp. 167-184 in *Measurement Errors in Surveys*, P. Biemer, R. Groves, L. Lyberg, N. Mathiowetz, and S. Sudman, eds. New York: Wiley.
Moon, Y.
  1998 Impression management in computer-based interviews: The effects of input modality, output modality, and distance. *Public Opinion Quarterly* 62:610-622.
Mosher, W.D., and A.P. Duffer, Jr.
  1994 Experiments in Survey Data Collection: The National Survey of Family Growth

Pretest. Paper presented at the annual meeting of the Population Association of America, Miami, FL.

Murphy, L.
1976 The Effects of the Attitude Supplement on NCS City Sample Victimization Data. Unpublished internal document, Bureau of the Census, Washington, DC.

National Victims Center
1992 *Rape in America: A Report to the Nation.* Arlington, VA: National Victims Center.

Neter, J., and J. Waksberg
1964 A study of response errors in expenditures data from household interviews. *Journal of the American Statistical Association* 59:17-55.

Phipps, P., and A. Tupek
1990 Assessing Measurement Errors in a Touchtone Recognition Survey. Paper presented at the International Conference on Measurement Errors in Surveys, Tucson, AZ.

Robinson, J.A.
1986 Temporal reference systems and autobiographical memory. Pp. 159-188 in *Autobiographical Memory*, D.C. Rubin, ed. Cambridge, England: Cambridge University Press.

Russell, D.E.H.
1982 The prevalence and incidence of forcible rape and attempted rape of females. *Victimology* 7:81-93.

Schaeffer, N.C.
1991 Conversation with a purpose—or conversation? Interaction in the standardized interview. Pp. 367-391 in *Measurement Errors in Surveys*, P.P. Biemer, R.M. Groves, L.E. Lyberg, N.A. Mathiowetz, and S. Sudman, eds. New York: Wiley.

Schober, M.
1999 Making sense of questions: An interactional approach. Pp. 77-93 in *Cognition and Survey Research*, M.G. Sirken, D.J. Herrmann, S. Schechter, N. Schwarz, J.M. Tanur, and R. Tourangeau, eds. New York: Wiley.

Schober, S., M.F. Caces, M. Pergamit, and L. Branden
1992 Effects of mode of administration on reporting of drug use in the National Longitudinal Survey. Pp. 267-276 in *Survey Measurement of Drug Use: Methodological Studies*, C. Turner, J. Lessler, and J. Gfroerer, eds. Rockville, MD: National Institute on Drug Abuse.

Schwarz, N., and G.L. Clore
1983 Mood, misattribution, and judgments of well-being: Informative and directive functions of affective states. *Journal of Personality and Social Psychology* 45:513-523.

Schwarz, N., H.J. Hippler, B. Deutsch, and F. Strack
1985 Response categories: Effects on behavioral reports and comparative judgments. *Public Opinion Quarterly* 49:388-395.

Schwarz, N., B. Knauper, H.J. Hippler, E. Noelle-Neumann, and F. Clark
1991 Rating scales: Numeric values may change the meaning of scale labels. *Public Opinion Quarterly* 55:618-630.

Schwarz, N., F. Strack, and H. Mai
  1991 Assimilation and contrast effects in part-whole question sequences: A conversational logic analysis. *Public Opinion Quarterly* 55:3-23.
Silver, B.D., P.R. Abramson, and B.A. Anderson
  1986 The presence of others and overreporting of voting in American national elections. *Public Opinion Quarterly* 50:228-239.
Sirken, M.G., D.J. Herrmann, S. Schechter, N. Schwarz, J. Tanur, and R. Tourangeau
  1999 *Cognition and Survey Research.* New York: Wiley.
Skogan, W.
  1981 *Issues in the Measurement of Victimization.* Washington, DC: Bureau of Justice Statistics.
Smith, T.W.
  1997 The impact of the presence of others on a respondent's answers to questions. *International Journal of Public Opinion Research* 9:33-47.
Strack, F., N. Schwarz, and M. Wänke
  1991 Semantic and pragmatic aspects of context effects in social and psychological research. *Social Cognition* 9:111-125.
Suchman, L., and B. Jordan
  1990 Interactional troubles in face-to-face survey interviews. *Journal of the American Statistical Association* 85:232-241.
Sudman, S., A. Finn, and L. Lannom
  1984 The use of bounded recall procedures in single interviews. *Public Opinion Quarterly* 48:520-524.
Sudman, S., N. Bradburn, and N. Schwarz
  1996 *Thinking About Answers: The Application of Cognitive Processes to Survey Methodology.* San Francisco: Jossey-Bass.
Tangney, J.P., R.W. Miller, L. Flicker, and D.H. Barlow
  1996 Are shame, guilt, and embarrassment distinct emotions? *Journal of Personality and Social Psychology* 70:1256-1269.
Thompson, C.P., J.J. Skowronski, S.F. Larsen, and A.L. Betz
  1996 *Autobiographical Memory.* Mahwah, NJ: Erlbaum.
Tourangeau, R.
  1999 Context effects on answers to attitude questions. Pp. 111-131 in *Cognition and Survey Research*, M.G. Sirken, D.J. Herrmann, S. Schechter, N. Schwarz, J. Tanur, and R. Tourangeau, eds. New York: Wiley.
Tourangeau, R., and T.W. Smith
  1996 Asking sensitive questions: The impact of data collection mode, question format, and question context. *Public Opinion Quarterly* 60:275-304.
Tourangeau, R., K. Rasinski, and N. Bradburn
  1991 Measuring happiness in surveys: A test of the subtraction hypothesis. *Public Opinion Quarterly* 55:255-266.
Tourangeau, R., L.J. Rips, and K.A. Rasinski
  2000 *The Psychology of Survey Response.* New York: Cambridge University Press.

Turner, C.F., J.T. Lessler, and J. Devore
  1992 Effects of mode of administration and wording on reporting of drug use. Pp. 177-220 in *Survey Measurement of Drug Use: Methodological Studies*, C. Turner, J. Lessler, and J. Gfroerer, eds. Rockville, MD: National Institute on Drug Abuse.

Turner, C.F., B.H. Forsyth, J.M. O'Reilly, P.C. Cooley, T.K. Smith, S.M. Rogers, and H.G. Miller
  1998 Automated self-interviewing and the survey measurement of sensitive behaviors. In *Computer-Assisted Survey Information Collection*, M.P. Couper, R.P. Baker, J. Bethlehem, C.Z.F. Clark, J. Martin, W.L. Nicholls II, and J. O'Reilly, eds. New York: Wiley.

Turner, C.F., L. Ku, S.M. Rogers, L.D. Lindberg, J.H. Pleck, and F.L. Sonenstein
  1998 Adolescent sexual behavior, drug use, and violence: Increased reporting with computer survey technology. *Science* 280:867-873.

Wagenaar, W.A.
  1986 My memory: A study of autobiographical memory over six years. *Cognitive Psychology* 18:225-252.

Williams, M.D., and J.D. Hollan
  1981 The process of retrieval from very long-term memory. *Cognitive Science* 5:87-119.

3

Comparison of Self-Report and Official Data for Measuring Crime

Terence P. Thornberry and Marvin D. Krohn

There are three basic ways to measure criminal behavior on a large scale. The oldest method is to rely on official data collected by criminal justice agencies, such as data on arrests or convictions. The other two rely on social surveys. In one case, individuals are asked if they have been victims of crime; in the other, they are asked to self-report their own criminal activity. This paper reviews the history of the third method—self-report surveys—assesses its validity and reliability, and compares results based on this approach to those based on official data. The role of the self-report method in the longitudinal study of criminal careers is also examined.

HISTORICAL OVERVIEW

The development and widespread use of the self-report method of collecting data on delinquent and criminal behavior together were one of the most important innovations in criminology research in the twentieth century. This method of data collection is used extensively both in the United States and abroad (Klein, 1989). Because of its common use, we often lose sight of the important impact that self-report studies have had on the study of the distribution and patterns of crime and delinquency, the etiological

---

This study was supported by the National Consortium on Violence Research.

study of criminality, and the study of the juvenile justice and criminal justice systems.

Sellin made the simple but critically important observation that "the value of a crime rate for index purposes decreases as the distance from the crime itself in terms of procedure increases" (1931:337). Thus, prison data are less useful than court or police data as a measure of actual delinquent or criminal behavior. Moreover, the reactions of the juvenile and criminal justice systems often rely on information from victims or witnesses of crime. It does not take an expert on crime to recognize that a substantial amount of crime is not reported and, if reported, is not officially recorded. Thus, reliance on official sources introduces a number of layers of potential bias between the actual behavior and the data. Yet, through the first half of the twentieth century, our understanding of the behavior of criminals and those who reacted to crime was based almost entirely on official data.

While researchers were aware of many of these limitations, the dilemma they faced was how to obtain valid information on crime that was closer to the source of the behavior. Observing the behavior taking place would be one method of doing so, but given the illegal nature of the behavior and the potential consequences if caught committing the behavior, participants in crime are reluctant to have their behavior observed. Even when observational studies have been conducted—for example, gang studies (e.g., Thrasher, 1927)—researchers could observe only a very small portion of the crimes that took place. Hence, observational studies had limited utility in describing the distribution and patterns of criminal behavior.

If one could not observe the behavior taking place, self-reports of delinquent and criminal behavior would be the data source nearest to the actual behavior. There was great skepticism, however, about whether respondents would be willing to tell researchers about their participation in illegal behaviors. Early studies (Porterfield, 1943; Wallerstein and Wylie, 1947) found that not only were respondents willing to self-report their delinquency and criminal behavior, they did so in surprising numbers.

Since those very early studies, the self-report methodology has become much more sophisticated in design, making it more reliable and valid and extending its applicability to myriad issues. Much work has been done to improve the reliability and validity of self-reports, including the introduction of specialized techniques intended to enhance the quality of self-report data. These developments have made self-report studies an integral part of the way crime and delinquency are studied.

Although the self-report method began with the contributions of

Porterfield (1943, 1946) and Wallerstein and Wylie (1947), the work of Short and Nye (1957, 1958) "revolutionized ideas about the feasibility of using survey procedures with a hitherto taboo topic" and changed how the discipline thought about delinquent behavior itself (Hindelang et al., 1981: 23). Short and Nye's research is distinguished from previous self-report measures in their attention to methodological issues, such as scale construction, reliability and validity, and sampling and their explicit focus on the substantive relationship between social class and delinquent behavior. A 21-item list of criminal and antisocial behaviors was used to measure delinquency, although in most of their analyses a scale comprised of a subset of only seven items was employed. Focusing on the relationship between delinquent behavior and the socioeconomic status of the adolescents' parents, Nye et al. (1958) found that relatively few of the differences in delinquent behavior among the different socioeconomic status groups were statistically significant.

Short and Nye's work stimulated much interest in both use of the self-report methodology and the relationship between some measure of social status (socioeconomic status, ethnicity, race) and delinquent behavior. The failure to find a relationship between social status and delinquency served at once to question extant theories built on the assumption that an inverse relationship did in fact exist and to suggest that the juvenile justice system may be using extra-legal factors in making decisions concerning juveniles who misbehave. A number of studies in the late 1950s and early 1960s used self-reports to examine the relationship between social status and delinquent behavior (Akers, 1964; Clark and Wenninger, 1962; Dentler and Monroe, 1961; Empey and Erickson, 1966; Erickson and Empey, 1963; Gold, 1966; Reiss and Rhodes, 1959; Slocum and Stone, 1963; Vaz, 1966; Voss, 1966). These studies advanced the use of the self-report method by applying it to different, more ethnically diverse populations (Clark and Wenninger, 1962; Gold, 1966; Voss, 1966), attending to issues concerning validity and reliability (Clark and Tifft, 1966; Dentler and Monroe, 1961; Gold, 1966), and constructing measures of delinquency that specifically addressed issues regarding offense seriousness and frequency (Gold, 1966). These studies found that, while most juveniles engaged in some delinquency, relatively few committed serious delinquency repetitively. With few exceptions, these studies supported the general conclusion that, if there were any statistically significant relationship between measures of social status and self-reported delinquent behavior, it was weak and clearly did not mirror the findings of studies using official data sources.

During this period of time researchers began to recognize the true potential of the self-report methodology. By including questions concerning other aspects of an adolescent's life as well as a delinquency scale on the same questionnaire, researchers could explore a host of etiological issues. Theoretically interesting issues concerning the family (Dentler and Monroe, 1961; Gold, 1970; Nye et al., 1958; Stanfield, 1966; Voss, 1964), peers (Erickson and Empey, 1963; Gold, 1970; Matthews, 1968; Reiss and Rhodes, 1964; Short, 1957; Voss, 1964), and school (Elliott, 1966; Gold, 1970; Kelly, 1974; Polk, 1969; Reiss and Rhodes, 1963) emerged as the central focus of self-report studies. The potential of the self-report methodology in examining etiological theories of delinquency was perhaps best displayed in Hirschi's (1969) *Causes of Delinquency*.

The use of self-report studies to examine theoretical issues continued throughout the 1970s. In addition to several partial replications of Hirschi's arguments (Conger, 1976; Hepburn, 1976; Hindelang, 1973; Jensen and Eve, 1976), other theoretical perspectives such as social learning theory (Akers et al., 1979), self-concept theory (Jensen, 1973; Kaplan, 1972), strain theory (Elliott and Voss, 1974; Johnson, 1979), and deterrence theory (Anderson et al., 1977; Jensen et al., 1978; Silberman, 1976; Waldo and Chiricos, 1972) were evaluated using data from self-report surveys.

Another development during this period was the introduction of national surveys on delinquency and drug use. Williams and Gold (1972) conducted the first nationwide survey, with a probability sample of 847 boys and girls 13 to 16 years old. Monitoring the Future (Johnston et al., 1996) is a national survey on drug use that has been conducted annually since 1975. It began as an in-school survey of a nationally representative sample of high school seniors and was expanded to include eighth- and tenth-grade students.

One of the larger undertakings on a national level is the National Youth Survey (NYS), conducted by Elliott and colleagues (1985). The NYS began in 1976 by surveying a national probability sample of 1,725 youth ages 11 through 17. The survey design was sensitive to a number of methodological deficiencies of prior self-report studies and has been greatly instrumental in improving the self-report method. The NYS is also noteworthy because it is a panel design, having followed the original respondents into their thirties.

Despite the expanding applications of this methodology, questions remained about what self-report instruments measure. The discrepancy in findings regarding the relationship between social status and delinquency based on self-report data versus official (and victim) data continued to perplex scholars. Early on, self-reports came under heavy criticism on a number of counts, including the selection of respondents and the selection of delinquency items. Nettler (1978:98) stated that "an evaluation of these unofficial ways of counting crime does not fulfill the promise that they would provide a better enumeration of offensive activity." Gibbons (1979:84) was even more critical in his summary evaluation, stating:

> The burst of energy devoted to self-report studies of delinquency has apparently been exhausted. This work constituted a criminological fad that has waned, probably because such studies have not fulfilled their early promise.

Two studies were particularly instrumental at that time in pointing to flaws in self-report measures. Hindelang and colleagues (1979) illustrated the problems encountered when comparing the results from studies using self-reports and those using official data or victimization data by comparing characteristics of offenders across the three data sources. They observed more similarity in those characteristics between victimization and Uniform Crime Reports data than between self-report data and the other two sources. They argued that self-report instruments did not include the more serious crimes for which people are arrested and that are included in victimization surveys. Thus, self-reports tap a different, less serious domain of behaviors than either of the other two sources, and discrepancies in observed relationships when using self-reports should not be surprising. The differential domain of crime tapped by early self-report measures could also explain the discrepancy in findings regarding the association between social status and delinquency.

Elliott and Ageton (1980) also explored the methodological shortcomings of self-reports. They observed that a relatively small number of youth commit a disproportionate number of serious offenses. However, most early self-report instruments failed to include serious offenses in the inventory and truncated the response categories for the frequency of offenses. In addition, many of the samples did not include enough high-rate offenders to clearly distinguish them from other delinquents. By allowing respondents to report the number of delinquent acts they committed rather than specifying an upper limit (e.g., 10 or more) and by focusing on high-rate offenders, Elliott and Ageton found relationships between engaging in serious delinquent behavior and race and social class that are more consistent with results from studies using official data.

Hindelang and colleagues (1979) and Elliott and Ageton (1980) suggested designing self-report studies so that they would acquire sufficient data from those high-rate, serious offenders who would be most likely to come to the attention of the authorities. They also suggested a number of changes in the way in which self-report data are measured, so that the data reflect the fact that some offenders contribute disproportionately to the rate of serious and violent delinquent acts.

The development of instruments to better measure serious offenses and the suggestion to acquire data from high-rate offenders coincided with a substantive change in the 1980s in the focus of much criminology work on the etiology of offenders. The identification of a relatively small group of offenders who commit a disproportionate amount of crime and delinquency led for a call to focus research efforts on "chronic" or "career" criminals (Blumstein et al., 1986; Wolfgang et al., 1972, 1987). Blumstein et al.'s observation that we need to study the careers of criminals, including early precursors of delinquency, maintenance through the adolescent years, and later consequences during the adult years, was particularly important in recognizing the need for examining the life-course development of high-rate offenders with self-report methodology.

The self-report methodology continues to advance in terms of both its application to new substantive areas and the improvement of its design. Gibbons's (1979) suggestion that self-reports were just a fad, likely to disappear, is clearly wrong. Rather, with improvements in question design, administration technique, reliability and validity, and sample selection, this technique is being used in the most innovative research on crime and delinquency. The sections that follow describe the key methodological developments that have made such applications possible.

DEVELOPMENT OF THE SELF-REPORT METHOD

Self-report measures of delinquent behavior have advanced remarkably in the 30-odd years since their introduction by Short and Nye (1957). Considerable attention has been paid to the development and improvement of their psychometric properties. The most sophisticated and influential work was done by Elliott and colleagues (Elliott and Ageton, 1980; Elliott et al., 1985; Huizinga and Elliott, 1986) and by Hindelang, Hirschi, and Weis (1979, 1981). From their work a set of characteristics for acceptable (i.e., reasonably valid and reliable) self-report scales has emerged. Five of the most salient of these characteristics are the inclusion of (1) a wide array of offenses, including serious offenses; (2) frequency response sets; (3)

screening for trivial behaviors; (4) application to a wider age range; and (5) the use of longitudinal designs. Each is discussed below.

- *Inclusion of a wide array of delinquency items.* The domain of crime covers a wide range of behaviors, from petty theft to aggravated assault and homicide. If the general domain of delinquent and criminal behavior is to be represented in a self-report scale, it is necessary for the scale to cover that same wide array of human activity. Simply asking about a handful of these behaviors does not accurately represent the theoretical construct of crime. In addition, empirical evidence suggests that crime does not have a clear unidimensional structure that would facilitate the sampling of a small number of items from a theoretically large pool to adequately represent the entire domain.

These considerations suggest that an adequate self-report scale for delinquency will be relatively lengthy. Many individual items are required to represent the entire domain of delinquent behavior, to represent each of its subdomains, and to ensure that each subdomain (e.g., violence, drug use) is itself adequately represented.

In particular, it is essential that a general self-reported delinquency scale tap serious as well as less serious behaviors. Early self-report scales tended to ignore serious criminal and delinquent events and concentrated almost exclusively on minor forms of delinquency. Failure to include serious offenses misrepresents the domain of delinquency and contaminates comparisons with other data sources. In addition, it misrepresents the dependent variable of many delinquency theories (e.g., Elliott et al., 1985; Thornberry, 1987) that attempt to explain serious, repetitive delinquency.

- *Inclusion of frequency response sets.* Many early self-report studies relied on response sets with a relatively small number of categories, thus tending to censor high-frequency responses. For example, Short and Nye (1957) used a four-point response with the highest category being "often." Aggregated over many items, these limited response sets had the consequence of lumping together occasional and high-rate delinquents, rather than discriminating between these behaviorally different groups.
- *Screening for trivial behaviors.* Self-report questions have a tendency to elicit reports of trivial acts that are very unlikely to elicit official reactions and even acts that are not violations of the law. This occurs more frequently with the less serious offenses but also plagues responses to serious

offenses. For example, respondents have included as thefts such pranks as hiding a classmate's books in the respondent's locker between classes, or as serious assault events that are really roughhousing between siblings.

Some effort must be made to adjust or censor the data to remove these events if the delinquency of the subjects is to be reflected properly and if the rank order of subjects with respect to delinquency is to be portrayed properly. Two strategies are generally available. First, one can ask a series of follow-up questions designed to elicit more information about an event, such as the value of stolen property, the extent of injury to the victim, and the like. Second, one can use an open-ended question asking the respondent to describe the event and then probe to obtain the information necessary to classify the act. Both strategies have been used with some success.

- *Application to a wider age range.* With increasing emphasis on the study of crime across the entire life course, self-report surveys have had to be developed to take into account both the deviant behavior of very young children and the criminal behavior of older adults. The behavioral manifestations of illegal behaviors or the precursors of such behavior can change depending on the stage in the life course at which the assessment takes place. For the very young child, measures have been developed that are administered to parents to assess antisocial behavior such as noncompliance, disobedience, and aggression (Achenbach, 1992). For the school-age child, Loeber and colleagues (1993) have developed a checklist that expands the range of antisocial behaviors to include such behaviors as stubbornness, lying, bullying, and other externalizing problems.

There has been less development of instruments targeted at adults. Weitekamp (1989) has criticized self-report studies for being primarily concerned with the adolescent years and simply using the same items for adults. This is particularly crucial given the concern over the small but very significant problem of chronic violent offenders.

- *Use of longitudinal designs.* Perhaps the most significant development in the application of the self-report methodology is its use in following the same subjects over time in order to account for changes in their criminal behavior. This has enabled researchers to examine the effect of age of onset, to track the careers of offenders, to study desistance, and to apply developmental theories to study both the causes and consequences of criminal behavior over the life course.

While broadening the range of issues that can be examined, application of the self-report technique within longitudinal panel designs introduces potential threats to the reliability and validity of the data. In addition to concern over construct continuity in applying the technique to different-aged respondents, researchers need to consider the possibility of panel or testing effects.

All of these newer procedures are likely to improve the validity, and to some extent the reliability, of self-report scales since they improve our ability to identify delinquents and to discriminate among different types of delinquents. These are clearly desirable qualities.

To gain these desirable qualities, however, requires a considerable expansion of the self-report schedule. This can be illustrated by describing the major components of the index currently being used in the Rochester Youth Development Study (Thornberry et al., in press) as well as the other two projects of the Program of Research on the Causes and Correlates of Delinquency (see Browning et al., 1999). The inventory includes 32 items that tap general delinquency and 12 that tap drug use, for a total of 44 items. For each item the subjects are asked if they committed the act since the previous interview. For all items to which they respond in the affirmative, a series of follow-up questions are asked, such as whether they had been arrested. In addition, for the most serious instance of each type of delinquency reported in the past six months, subjects are asked to describe the event by responding to the question: "Could you tell me what you did?" If that open-ended question does not elicit the information needed to describe the event adequately, a series of questions, which vary from 2 to 14 probes depending on the offense, are asked.

Although most of these specific questions are skipped for most subjects since delinquency remains a rare event, this approach to measuring self-reported delinquency is a far cry from the initial days of the method, when subjects used a few categories to respond to a small number of trivial delinquencies with no follow-up items. Below we evaluate the adequacy of this approach for measuring delinquent and criminal behavior.

RELIABILITY AND VALIDITY

For any measure to be scientifically worthwhile it must possess both reliability and validity. Reliability is the extent to which a measuring procedure yields the same result on repeated trials. No measure is absolutely, perfectly reliable. Repeated use of a measuring instrument will always produce some variation from one application to another. That variation can be very slight or quite large. So the central question in assessing the reliability of a measure is not whether it is reliable but how reliable it is; reliability is always a matter of degree.

Validity is a more abstract notion. A measure is valid to the extent to which it measures the concept you set out to measure and nothing else. While reliability focuses on a particular property of the measure—namely, its stability over repeated uses—validity concerns the crucial relationship between the theoretical concept one is attempting to measure and what one actually measures. As is true with the case of reliability, the assessment of validity is not an either/or proposition. There are no perfectly valid measures, but some measures are more valid than others. We now turn to an assessment of whether self-reported measures of delinquency are psychometrically acceptable.

Assessing Reliability

There are two classic ways to assess the reliability of social science measures: test-retest reliability and internal consistency. Huizinga and Elliott (1986) make a convincing case that the test-retest approach is fundamentally more appropriate for assessing self-reported measures of delinquency.

Internal consistency means that multiple items measuring the same underlying concept should be highly intercorrelated. Although a reasonable expectation for attitudinal measures, this expectation is less reasonable for behavioral inventories such as self-report measures of delinquency. Current self-report measures typically include 30 to 40 items measuring a wide array of delinquent acts. Just because someone was truant is no reason to expect that they would be involved in theft or vandalism. Similarly, if someone reports that they have been involved in assaultive behavior, there is no reason to assume they have been involved in drug sales or loitering. Indeed, given the relative rarity of involvement in delinquent acts, it is very likely that most people will respond in the negative to most items and in the affirmative to only a few items. This is especially the case if we are asking about short reference periods (e.g., the past year or past six months). There is no strong underlying expectation that the responses will be highly intercorrelated, and therefore an internal consistency approach to assessing reliability may not be particularly appropriate. (See Huizinga and Elliott, 1986, for a more formal discussion of this point.)

Some theories of crime (e.g., Gottfredson and Hirschi, 1990; Jessor et al., 1991) assume there is an underlying construct, such as self-control, that generates versatility in offending. If so, there should be high internal consistency among self-reported delinquency items. While this result may be supportive of the theoretical assumption, it is not necessarily a good indicator of the reliability of the measures. If internal consistency were low, it may not have any implication for reliability but may simply mean that this particular theoretical assumption was incorrect. Nevertheless, we do note that studies that have examined the internal consistency of self-report measures generally find acceptable alpha coefficients. For example, Hindelang and colleagues report alphas between 0.76 and 0.93 for various self-report measures (1981:80).

We will focus our attention on the test-retest method of assessing reliability. This approach is quite straightforward. A sample of respondents is administered a self-reported delinquency inventory (the test) and then, after a short interval, the same inventory is readministered (the retest). In doing this the same questions and the same reference period should be used at both times.

The time lag between the test and the retest is also important. If it is too short, it is likely the answers provided on the retest will be a function of memory. If so, estimates of reliability would be inflated. If the time period between the test and the retest is too great, it is likely the responses given on the retest would be less accurate than those given on the test because of memory decay. In this case the reliability of the scale would be underestimated. There is no hard-and-fast rule for assessing the appropriateness of this lag, but somewhere in the range of one to four weeks appears to be optimal.

A number of studies have assessed the test-retest reliability of self-reported delinquency measures. In general, the results indicate that these measures are acceptably reliable. The reliability coefficients vary somewhat depending on the number and types of delinquent acts included in the index and the scoring procedures used (e.g., simple frequencies or ever-variety scores), but scores well above 0.80 are common. In summarizing much of the previous literature in this area, Huizinga and Elliott (1986:300) state:

> Test-retest reliabilities in the 0.85 - 0.99 range were reported by several studies employing various scoring schemes and numbers of items and using test-retest intervals of from less than one hour to over two months (Kulik et al., 1968; Belson, 1968; Hindelang et al., 1981; Braukmann et al., 1979;

Patterson and Loeber, 1982; Skolnick et al., 1981; Clark and Tifft, 1966; Broder and Zimmerman, 1978).

Perhaps the most comprehensive assessment of the psychometric properties of the self-report method was conducted by Hindelang and his colleagues (1981). Their self-report inventory was quite extensive, consisting of 69 items divided into the following major subindices: official contact index, serious crime index, delinquency index, drug index, and school and family offenses index. While mindful of the limitations of internal consistency approaches, Hindelang and colleagues (1981) reported Cronbach's alpha coefficients for a variety of demographic subgroups and for the ever-variety, last-year variety, and last-year frequency scores. The coefficients range from 0.76 to 0.93. Most of the coefficients are above 0.8, and 8 of the 18 coefficients are above 0.9.

Hindelang and colleagues (1981) also estimated test-retest reliabilities for these three self-report measures for each of the demographic subgroups. Unfortunately, only 45 minutes elapsed between the test and the retest, so it is quite possible the retest responses were strongly influenced by memory effects. Nevertheless, most of the test-retest correlations are above 0.9.

Hindelang et al. point out that reliability scores of this magnitude are higher than those typically associated with many attitudinal measures and conclude that "the overall implication is that in many of the relations examined by researchers, the delinquency dimension is more reliably measured than are many of the attitudinal dimensions studied in the research" (p. 82).

The other major assessment of the psychometric properties of the self-report method was conducted by Huizinga and Elliott (1986) using data from the NYS. At the fifth NYS interview, 177 respondents were randomly selected and reinterviewed approximately four weeks after their initial assessment. Based on these data, Huizinga and Elliott estimated test-retest reliability scores for the general delinquency index and a number of subindices. They also estimated reliability coefficients for frequency scores and variety scores.

The general delinquency index appears to have an acceptable level of reliability. The test-retest correlation for the frequency score is 0.75 and for the variety score, 0.84. For the various subindices—ranging from public disorder offenses to the much more serious index offenses—the reliabilities vary from a low of 0.52 (for the frequency measure of felony theft) to a high of 0.93 (for the frequency measure of illegal services). In total,

Huizinga and Elliott report 22 estimates of test-retest reliability (across indices and across frequency and variety scores), and the mean reliability coefficient is 0.74.

Another way Huizinga and Elliott assessed the level of test-retest reliability is by estimating the percentage of the sample that changed frequency responses by two or less. If the measure is highly reliable, few changes would be expected over time. For most subindices there appears to be acceptable reliability based on this measure. For example, for index offenses 97 percent of respondents changed their answers by two delinquent acts or less. Huizinga and Elliott (1986:303) summarized these results as follows:

> Scales representing more serious, less frequently occurring offenses (index offenses, felony assault, felony theft, robbery) have the highest precision, with 96 to 100 percent agreement, followed by the less serious offenses (minor assault, minor theft, property damage), with 80 to 95 percent agreement. The public disorder and status scales have lower reliabilities (in the 40 to 70 percent agreement range), followed finally by the general SRD [self-reported delinquency] scale, which, being a composite of the other scales, not surprisingly has the lowest test-retest agreement.

Huizinga and Elliott did not find any consistent differences across sex, race, class, place of residence, or delinquency level in terms of test-retest reliabilities (see also Huizinga and Elliott, 1983).

Assessing Validity

There are several ways to assess validity. We concentrate on three: content validity, construct validity, and criterion validity.

Content Validity

Content validity is a subjective or logical assessment of the extent to which a measure adequately reflects the full domain, or full content, that is contained in the concept being measured. To argue that a measure has content validity, the following three criteria must be met. First, the domain of the concept must be defined clearly and fully. Second, questions or items must be created that cover the whole range of the concept under investigation. Third, items or questions must be sampled from that range so that the ones that appear on the test are representative of the underlying concept.

A reasonable definition of delinquency and crime is the commission of behaviors that violate criminal law and that place the individual at some risk of arrest if the behavior were known to the police. Can a logical case be made that self-report measures of delinquency are valid in this respect?

As noted above, the earlier self-report inventories contained relatively few items to measure the full range of delinquent behaviors. For example, Short and Nye's (1957) inventory contains only 21 items, and most of their analysis was conducted with a 7-item index. Similarly, Hirschi's self-report measure (1969) is based on only 6 items. More importantly, the items included in these scales are clearly biased toward the minor or trivial end of the continuum.

The more recent self-report measures appear to be much better in this regard. For example, the Hindelang and colleagues (1981) index includes 69 items that range from status offenses, such as skipping class, to violent crimes, like serious assault and armed robbery. The NYS index (Elliott et al., 1985) has 47 items designed to measure all but one (homicide) of the Uniform Crime Reports Part I offenses, 60 percent of the Part II offenses, and offenses that juveniles are particularly likely to commit. The self-report inventory used by the three projects of the Program of Research on the Causes and Correlates of Delinquency has 32 items that measure delinquent behavior and 12 that measure substance use.

These more recent measures, while not perfect, tap into a much broader range of delinquent and criminal behaviors. As a result, they appear to have reasonable content validity.

Construct Validity

Construct validity refers to the extent to which the measure being validated is related in theoretically expected ways to other concepts or constructs. In our case the key question is: Are measures of delinquency based on the self-report method correlated in expected ways with variables expected to be risk factors for delinquency?

In general, self-report measures of delinquency and crime, especially the more recent longer inventories, appear to have a high degree of construct validity. They are generally related in theoretically expected ways to basic demographic characteristics and to a host of theoretical variables drawn from various domains such as individual attributes, family structure and processes, school performance, peer relationships, neighborhood characteristics, and so forth. Hindelang and colleagues (1981) offer one of the clearer assessments of construct validity. They correlate a number of etiological variables with different self-report measures collected under different conditions. With a few nonsystematic exceptions, the correlations are in the expected direction and of the expected magnitude.

Overall, construct validity may offer the strongest evidence for the validity of self-report measures of delinquency and crime. Indeed, if one examines the general literature on delinquent and criminal behavior, it is surprising how few theoretically expected relationships are not observed for self-reported measures of delinquency and crime. It is unfortunate that this approach is not used to assess validity more formally and more systematically.

Criterion Validity for Delinquency and Crime

Criterion validity "refers to the relationship between test scores and some known external criterion that adequately indicates the quantity being measured" (Huizinga and Elliott, 1986: 308). There is a fundamental difficulty in assessing the criterion validity of self-reported measures of delinquency and crime and for that matter all measures of delinquency and crime. Namely, there is no gold standard by which to judge the self-report measure. That is, there is no fully accurate assessment that can be used as a benchmark. In contrast, to test the validity of self-reports of weight, people could be asked to self-report their weight and each respondent could then be weighed on an accurate scale—the external criterion. Given the secretive nature of criminal behavior, however, there is nothing comparable to a scale in the world of crime. As a result, the best that can be done is to compare different flawed measures of criminal involvement to see if there are similar responses and results. If so, the similarity across different measurement strategies heightens the probability that the various measures are tapping into the underlying concept of interest. While not ideal, this is the best that can be done in this area of inquiry.

There are several ways to assess criterion validity. One of the simplest is called known group validity. In this approach one compares scores for groups of people who are likely to differ in terms of their underlying involvement in delinquency. For example, the delinquency scores of seminarians would be expected to be lower than the delinquency scores of street gang members.

Over the years a variety of group comparisons have been made to assess the validity of self-report measures. They include comparisons between individuals with and without official arrest records, between individuals convicted and not convicted of criminal offenses, and between institutionalized adolescents and high school students. In all cases these comparisons indicate that the group officially involved with the juvenile justice system self-reports substantially more delinquent acts than the other group. (See, for example, the work by Erickson and Empey, 1963; Farrington, 1973; Hardt and Petersen-Hardt, 1977; Hindelang et al., 1981; Hirschi, 1969; Kulik et al., 1968; Short and Nye, 1957; and Voss, 1963.)

While comparisons across known groups are helpful, they offer a minimal test of criterion validity. The real issue is not whether groups differ but the extent to which *individuals* have similar scores on the self-report measure and on other measures of criminal behavior. A variety of external criteria have been used (see the discussion in Hindelang et al., 1981). The two most common approaches are to compare self-reported delinquency scores with official arrest records and self-reports of arrest records with official arrest records.

We can begin by examining the correlation between self-reported official contacts and official measures of delinquency. These correlations are quite high in the Hindelang et al. study, ranging from 0.70 to 0.83. Correlations of this magnitude are reasonably large for this type of data.[1]

The most recent investigation of this topic is by Maxfield, Weiler, and Widom (2000), using Widom's (1989) sample of child maltreatment victims and their matched controls. Unlike most studies in this area, the respondents were adults (mean age = 28). They were interviewed only once, so all of the self-reported arrest data are retrospective, with relatively long recall periods. Nevertheless, the concordance between having an official arrest and a self-report of being arrested is high. Of those arrested, 73 percent reported an arrest. Maxfield et al. noted lower levels of reporting for females than males and for blacks than whites. The gender differences were quite persistent, but the race differences were more pronounced for less frequent offenders and diminished considerably for more frequent offenders.

Maxfield et al. also studied "positive bias," the self-reporting of arrests that are not found in official records. They found that 21 percent of respondents with no arrest history self-reported being arrested. Positive bias

---

[1]This is particularly the case given the level of reliability that self-reported data have (see previous section). By adding random error to the picture, poor reliability attenuates or reduces the size of the observed correlation coefficients.

was higher for males than females, but there were no race differences. It is not clear whether this is a problem with the self-reports (i.e., positive bias) or with the official records such as sealed records, sloppy record keeping, use of aliases, and so forth. This is an understudied topic that needs greater investigation.

The generally high level of concordance between self-reports of being arrested or having a police contact and having an official record has been observed in other studies as well (Hardt and Petersen-Hardt, 1977; Hathaway et al., 1960; Rojek, 1983). When convictions are examined, even higher concordance rates are reported (Blackmore, 1974; and Farrington, 1973).

It appears that survey respondents are quite willing to self-report their involvement with the juvenile justice and criminal justice systems. Are they also willing to self-report their involvement in undetected delinquent behavior? This is the central question. The best way to examine this is to compare self-reported delinquent behavior and official measures of delinquency. If these measures are valid, a reasonably large positive correlation between them would be expected.

Hindelang and colleagues (1981) presented correlations using a number of different techniques for scoring the self-report measures, but here we focus on the average correlation across these different measures and on the correlation based on the ever-variety scores, as presented in their Figure 3. Overall, these correlations are reasonably high, somewhere around 0.60 for all subjects. The most important data though are presented for race-by-gender groups. For white and African American females and for white males, the correlations range from 0.58 to 0.65 when the ever-variety score is used; for the correlations that are averaged across the different self-report measures, the magnitudes range from 0.50 to 0.60. For African American males, however, the correlation is at best moderate. For the ever-variety self-reported delinquency score, the correlation is 0.35, and the average across the other self-reported measures is 0.30.

Huizinga and Elliott (1986), using data from the NYS, also examined the correspondence between self-reports of delinquent behavior and official criminal histories. They recognized that there can be considerable slippage between these two sources of data even when the same event is actually contained in both data sets. For example, while an adolescent can self-report a gang fight, it may be recorded in the arrest file as disturbing the peace, or an arrest for armed robbery can be self-categorized as a mugging or theft by the individual. Because of this, Huizinga and Elliott provided two levels of matching, "tight matches" and "broad matches." The analysis provides information on both the percentage of people who provide tight and broad matches to their arrest records and the percentage of arrests that are matched by self-reported behavior.

For the tight matches, almost half of the respondents (48 percent) concealed or forgot at least some of their offensive behavior, and about a third (32 percent) of all the offenses were not reported. When the broad matches are used, the percentage of respondents concealing or forgetting some of their offenses dropped to 36 percent and the percentage of offenses not self-reported to 22 percent. While the rates of underreporting are substantial, it should be noted that the majority of individuals who have been arrested self-report their delinquent behavior, and the majority of offenses they commit also are reported.

The reporting rates for gender, race, and social class groupings are quite comparable to the overall rates, with one exception. As was the case with the Seattle data, African American males substantially underreported their involvement in delinquency.

Farrington and colleagues (1996), using data from the middle and oldest cohorts of the Pittsburgh Youth Study, also examined this issue. The Pittsburgh study, as one of three projects in the Program of Research on the Causes and Correlates of Delinquency, uses the same self-reported delinquency index as described earlier for the Rochester Youth Development Study. Farrington et al. classified each of the boys in the Pittsburgh study into one of four categories based on the seriousness of their self-reported delinquency: no delinquency, minor delinquency only, moderate delinquency only, and serious delinquency. They then used juvenile court petitions as an external criterion to assess the validity of the self-reported responses. Both concurrent validity and predictive validity were assessed.

Overall, this analysis suggests that there is a substantial degree of criterion validity for the self-report inventory used in the Program of Research on the Causes and Correlates of Delinquency. Respondents who are in the most serious category based on their self-report responses are significantly more likely to have juvenile court petitions, both concurrently and predictively. For example, the odds ratio of having a court petition for delinquency is about 3:0 for the respondents in the most serious self-reported delinquency category versus the other three.

African American males are no more or less likely to self-report delinquent behavior than white males. With few exceptions, the odds ratios comparing self-reported measures and official court petitions are significant for both African Americans and whites; in some cases the odds ratios are higher for whites, and in some cases they are higher for African Americans.

These researchers also compared the extent to which boys with official court petitions self-reported being apprehended by the police. Overall, about two-thirds of the boys with court petitions answered in the affirmative. Moreover, there was no evidence of differential validity. Indeed, the African American respondents were more likely to admit being apprehended by the police than were the white respondents. Farrington and his colleagues (1996:509) concluded that "concurrent validity for admitting offenses was higher for Caucasians but concurrent validity for admitting arrests was higher for African Americans. There were no consistent ethnic differences in predictive validity."

Finally, Farrington and colleagues (2000) used data from the Seattle Social Development Project to assess the concurrent and predictive validity of self-report data. They compared self-report responses for a variety of indices and offense types to the odds of having a court referral. For the general delinquency index the concurrent odds ratio was 2:8 and the predictive odds ratio was 2:2. Validity was highest for self-reports of drug involvement and lowest for property offenses, with violent offenses falling in the middle.

Putting all this together leads to a somewhat mixed assessment of the validity of self-report measures. On the one hand, it seems that the overall validity of self-report data is in the moderate-to-strong range, especially for self-reports of being arrested. For the link between self-reported delinquent behavior and official measures of delinquency, the only link based on independent sources of data, the overall correlations are somewhat smaller but still quite acceptable. On the other hand, looking at the issue of differential validity, there are some disturbing differences by race. It is hard to determine whether this is a problem with the self-report measures, the official measures, or both. We will return to a discussion of this issue after additional data are presented.

Criterion Validity for Substance Use

The previous studies focused on delinquent or criminal behavior where, as mentioned earlier, there is no true external criterion for evaluating validity. There is an external criterion for one class of deviant behavior—substance use. Physiological data (e.g., from saliva or urine) can be used to independently assess recent use of various substances. The physiological data can then be compared to self-reports of substance use to assess the validity of the self-report instruments. A few examples of this approach can be offered.

We begin with a study of a minor form of deviant behavior—adolescent tobacco use. Akers and colleagues (1983) examined tobacco use among a sample of junior and senior high school students in Muscatine, Iowa. The respondents provided saliva samples that were used to detect nicotine use by the level of salivary thiocyanate. The students also self-reported whether they smoked and how often. The self-report data had very low levels of either underreporting of tobacco use or overreporting. Overall, the study estimated that 95 to 96 percent of the self-reported responses were accurate and valid.

The Drug Use Forecasting (DUF) project (1990), sponsored by the National Institute of Justice, is an ongoing assessment of the extensiveness of drug use for samples of arrestees in cities throughout the country. Individuals who have been arrested and brought to central booking stations are interviewed and asked to provide urine specimens. Both the urine samples and the interviews are provided voluntarily, and there is an 80 percent cooperation rate for the urine samples and a 90 percent cooperation rate for the interviews. The urine specimens are tested for 10 different drugs, and in some of the interviews there is a self-reported drug use inventory. Assuming the urine samples provide a reasonably accurate estimate of actual drug use, they can be used to validate self-reported information.

The results vary considerably by type of drug. There is generally a fairly high concordance for marijuana use. For example, in 1990 in New York City 28 percent of the arrestees self-reported marijuana use and 30 percent tested positive for marijuana use. Similarly, in Philadelphia 28 percent self-reported marijuana use and 32 percent tested positive. The worst comparison in this particular examination of the Drug Use Forecasting data was from Houston, where 15 percent of arrestees self-reported marijuana use and 43 percent tested positive.

For more serious drugs, the level of underreporting is much more severe. For example, 47 percent of the New York City arrestees self-reported cocaine use and 74 percent tested positive. Very similar numbers were generated in Philadelphia, where 41 percent self-reported cocaine use but 72 percent tested positive. Similar levels of underreporting were observed for other hard drugs such as heroin and in other cities.

The data collected in the Drug Use Forecasting project are obviously quite different from those collected in typical self-report surveys. The sample is limited to people who have just been arrested, and they are asked to provide self-incriminating evidence to a research team while in a central booking station. It is not entirely clear how this setting affects the results. On the one hand, individuals may be reluctant to provide additional self-incriminating evidence after having just been arrested. On the other hand, if one has just been arrested for a serious crime like robbery or burglary, admitting to recent drug use may not be considered a big deal. In any event, caution is needed in using these data to generalize to the validity of typical self-report inventories.

SUMMARY

We have examined three different approaches to assessing the validity of self-reported measures of delinquency and crime: content, construct, and criterion validity. Several conclusions, especially for the more recent self-report inventories, appear warranted.

The self-report method for measuring this rather sensitive topic—undetected criminal behavior—appears to be reasonably valid. The content validity of the recent inventories is acceptable, the construct validity is quite high, and the criterion validity appears to be in the moderate-to-strong range. Putting this all together, it could be concluded that for most analytical purposes, self-reported measures are acceptably accurate and valid.

Despite this general conclusion, there are still several substantial issues concerning the validity of self-report measures. First, the validity of the earlier self-report scales, and the results based on them, are at best questionable. Second, based on the results of the tests of criterion validity, there appears to be a substantial degree of either concealing or forgetting past criminal behavior. While the majority of individual respondents report their offenses and the majority of all offenses are reported, there is still a good deal of underreporting.

Third, there is an unresolved issue of differential validity. As compared to other race-gender groups, some studies have found that the responses provided by African American males appear to have lower levels of validity (Hindelang et al., 1981; Huizinga and Elliott, 1986). More recently, however, Farrington et al. (1996) and Maxfield et al. (2000) found no evidence of differential validity by race. Maxfield and colleagues (2000) did find lower reporting for females than males. The level of differential validity is one of the most important methodological issues confronting the self-report method and should be a high priority for future research efforts.

Fourth, based on studies of self-reported substance use, there is some evidence that validity may be less for more serious types of offenses. In the substance use studies, the concordance between the self-report and physiological measures was strongest for adolescent tobacco use, then for marijuana use, and it was weakest for hard drugs such as cocaine and heroin. A similar pattern is seen for several studies of self-reported delinquency and crime (e.g., Elliott and Voss, 1974; Huizinga and Elliott, 1986).

What then can be said about the psychometric properties of self-reported measures of delinquency and crime? With respect to reliability, this approach to measuring involvement in delinquency and crime appears to be acceptable. Most estimates of reliability are quite high, and there is no evidence of differential reliability. With respect to validity, the conclusion is a little murkier. There is a considerable amount of underreporting, and there is also the potential problem of differential validity. Nevertheless, content validity and construct validity appear to be quite high, and an overall estimate of criterion validity would be in the moderate-to-strong range. Perhaps the conclusion reached by Hindelang and colleagues (1981:114) is still the most reasonable:

> The self-report method appears to behave reasonably well when judged by standard criteria available to social scientists. By these criteria, the difficulties in self-report instruments currently in use would appear to be surmountable; the method of self-reports does not appear from these studies to be fundamentally flawed. Reliability measures are impressive and the majority of studies produce validity coefficients in the moderate to strong range.

SPECIALIZED RESPONSE TECHNIQUES

Because of the sensitive nature of this area—asking people to report previously undetected criminal behavior—there has always been concern about how best to ask such questions to maximize the accuracy of the responses. Some early self-report researchers favored self-administered questionnaires while others favored more personal face-to-face interviews. Similarly, some argued that anonymous responses were inherently better than nonanonymous responses. In their Seattle study, Hindelang and his colleagues (1981) directly tested these concerns by randomly assigning respondents to one of four conditions: nonanonymous questionnaire, anonymous questionnaire, nonanonymous interview, and anonymous interview.

Their results indicate that there is no strong method effect in producing self-report responses, and that no one approach is consistently better than the other approaches. Similar results were reported by Krohn and his colleagues (1974). Some research, especially in the alcohol and drug use area, has found a method effect. For example, Aquilino (1994) found that admission of alcohol and drug use is lowest in telephone interviews, somewhat higher in face-to-face interviews, and highest in self-administered questionnaires (see also Aquilino and LoSciuto, 1990; Turner et al., 1992). While evident, the effect size is typically not very large.

Although basic method effects do not appear to be very strong, there is still concern that in all of these approaches to the collection of survey data, respondents will feel vulnerable about reporting sensitive information. Because of that, a variety of more specialized techniques have been developed to protect the individual respondent's confidentiality, hopefully increasing the level of reporting.

Randomized Response Technique

The randomized response technique assumes that the basic problem with the validity of self-reported responses is that respondents are trying to conceal sensitive information; that is, they are unwilling to report undetected criminal behavior as long as there is any chance of others, including the researchers, linking the behavior to them. Randomized response techniques allow respondents to conceal what they really did while at the same time providing useful data to researchers. There are various ways to accomplish this, and how the basic process works can be illustrated with a simple example of measuring the prevalence of marijuana use. The basic question is: "Have you ever smoked marijuana?"

Imagine an interview setting in which there is a screen between the interviewer and respondent so that the interviewer cannot see what the respondent is doing. The interviewer asks a sensitive question (e.g., "Have you ever smoked marijuana?") with the following special instruction: Before answering, please flip a coin. If the coin lands on heads, please answer "yes" regardless of whether or not you smoked marijuana. If the coin lands on tails, please tell me the truth. It is impossible for the interviewer to know whether a "yes" response is produced by the coin or by the fact that the respondent actually smoked marijuana. In this way the respondent can admit to sensitive behavior but other people, including the interviewer, do not know if the admission is truthful or not.

From the resulting data the prevalence of marijuana use can be estimated quite easily. Say we receive 70 "yes" responses in a sample of 100 respondents. Fifty of those would be produced by the coin landing on heads, and these 50 respondents can simply be ignored. Of the remaining 50 respondents though, 20 said "yes" because they smoked marijuana, so the prevalence of marijuana use is 20 out of 50, or 40 percent.

This technique is not limited to "yes" or "no" questions or to flipping coins. Any random process can be used as long as the probability distribution of bogus versus truthful responses is known. From these data, prevalence, variety, and frequency scores and means and variances can be estimated, and the information can be correlated with other variables, just as is done with regular self-report data.

Weis and Van Alstyne (1979) tested a randomized response procedure in the Seattle study. They concluded that the randomized response approach is no more efficient in eliciting positive responses to sensitive items than are traditional methods of data collection. This finding is consistent with the overall conclusion in the Seattle study that the method of administration is relatively unimportant.

The other major assessment of the randomized response technique was conducted by Tracy and Fox (1981). They sampled people who had been arrested in Philadelphia and then went to their homes to interview them. Respondents were asked if they had been arrested and, if so, how many times. There were two methods of data collection: a randomized response procedure and a regular self-report interview.

The results indicated that the randomized response approach does make a difference. For all respondents there was about 10 percent less error in the randomized response technique. For respondents who had been arrested only once, the randomized response approach actually increased the level of error. But for recidivists the randomized response technique reduced the level of error by about 74 percent.

Also, the randomized response technique generated random errors; that is, the errors were not correlated with other important variables. The regular self-reported interview, however, generated systematic error or bias. In this approach, underreporting was related to females, African American females, respondents with a high need for approval, lower-income respondents, and persons with a larger number of arrests.

Overall, it is not clear to what extent a randomized response approach actually generates more complete and accurate reporting. The two major studies of this topic produced different results: Weis and Van Alstyne (1979) reported no effect, and Tracy and Fox (1981) reported sizable positive effects.

Computer-Assisted Interviewing

Advances in both computer hardware and software have made the introduction of computers in the actual data collection process not only a possibility but, according to Tourangeau and Smith (1996:276), "perhaps the most commonly used method of face-to-face data collection today." The use of computers in the data collection process began in the 1970s with computer-assisted telephone surveys (Saris, 1991). The technology was soon adapted to the personal interview setting with either the interviewer administering the schedule, the computer-assisted personal interview, or the respondent self-administering the schedule by reading the questions on the computer screen and entering the responses—the computer-assisted self-administered interview (CASI). It is also possible to have an audio version in which the questions are recorded and the respondent listens to them on headphones rather than having them read aloud by the interviewer. This is called an audio computer-assisted self-administered interview (ACASI).

One reason for the use of computer-assisted data collection that is particularly relevant for this paper is its potential for collecting sensitive information in a manner that increases the confidentiality of responses. Another advantage is that it allows for the incorporation of complex branching patterns (Beebe et al., 1998; Saris, 1991; Tourangeau and Smith, 1996; Wright et al., 1998). Computer software can be programmed to include skip patterns and increase the probability that the respondent will answer all appropriate questions. An added advantage of computer-assisted presentation is that the respondent does not see the implication of answering in the affirmative to questions with multiple follow-ups.

ACASI has two additional advantages. First, it circumvents the potential problem of literacy; the respondent does not have to read the questions. Second, in situations where other people might be nearby, the questions and responses are not heard by anyone but the respondent. Hence, the respondent can be more assured that answers to sensitive questions will remain private.

While computer-assisted administration of sensitive questions provides obvious advantages in terms of efficiency of presentation and data collection, the key question is the difference in the responses that are elicited when such technology is used. Tourangeau and Smith (1996) reviewed 18 studies that compared different modes of data collection. The types of behavior examined included health problems (e.g., gastrointestinal problems), sexual practices, abortion, and alcohol and drug use. Tourangeau and Smith indicate that techniques that are self-administered generally elicit higher rates of problematic behaviors than those administered by an interviewer. Moreover, CASIs elicit higher rates than either self-administered questionnaires or paper-and-pencil interviews administered by an interviewer. Also, ACASIs elicit higher rates than CASIs.

Estimates of prevalence rates of illegal and embarrassing behaviors appear to be higher when computer-assisted techniques, particularly those involving self-administration, are used. The higher prevalence rates need to be externally validated. The added benefits of providing for schedule complexity and consistency in responses make these techniques attractive, and it is clear that they will continue to be used with increasing frequency.

SELF-REPORT MEASURES ACROSS THE LIFE COURSE

One of the most exciting developments in criminology over the past 15 years has been the emergence of a life-course or developmental focus (Farrington, 1986; Jessor, 1998; Thornberry, 1997; Thornberry and Krohn, 2001; Weitekamp, 1989). Theoretical work has expanded from a narrow focus on the adolescent years to encompass the entire criminal careers of individuals, from the precursors of delinquency that are manifest in early childhood (Moffitt, 1997; Tremblay et al., 1999) through the high-delinquency years of middle and late adolescence, on into adulthood when most, but not all, offenders decrease their participation in illegal behaviors (Loeber et al., 1998; Moffitt, 1997; Sampson and Laub, 1990; Thornberry and Krohn, 2001). Research on criminal careers (Blumstein et al., 1986) has documented the importance of examining such issues as the age of onset (Krohn et al., 2001) and the duration of criminal activity (Wolfgang et al., 1987).

In addition, a growing body of research has demonstrated that antisocial behavior is rather stable from childhood to adulthood (Farrington, 1989a; Huesmann et al., 1984; Moffitt, 1993; Olweus, 1979). Much of this work has relied on official data. However, criminological research increasingly relies on longitudinal panel designs using self-report measures of antisocial behavior to understand the dynamics of criminal careers. Nevertheless, relatively little attention has been paid to the use of self-report techniques in longitudinal studies over the life course, even though this introduces a number of interesting measurement issues. Several of these issues are discussed in this section. Some of them involve the construction of valid measures at different developmental stages; others involve the consequences of repeated measures.

Construct Continuity

While many underlying theoretical constructs such as involvement in crime remain constant over time, their behavioral manifestations can change as subjects age. Failure to adapt measures to account for these changes may lead to age-inappropriate measures with reduced validity and reliability. To avoid this, measures need to adapt to the respondent's developmental stage to reflect accurately the theoretical constructs of interest (Campbell, 1990; LeBlanc, 1989; Patterson, 1993; Weitekamp, 1989). In some cases this may mean defining the concept at a level to accommodate the changing contexts in which people at different ages act. In other cases it may mean recognizing that different behaviors at different ages imply consistency in behavioral style (Campbell, 1990).

Construct continuity creates a difficult design dilemma. If the measure does not change to reflect the developmental stage, the accuracy of the measure is likely to deteriorate and the study of change is compromised. Changing the measure over time, however, creates its own set of problems, especially for the study of change. If change is observed, is it a function of changes in the person's behavior or of changes in the measure?

Relatively little attention has been paid to this issue in the study of criminal careers and, in particular, the study of self-report measures. At a more practical level, several studies have adapted self-report measures to both childhood and adulthood.

Self-Report Measures for Children

Antisocial behavior has been likened to a chimera (Patterson, 1993) with manifestations that change and accumulate with age. At very young ages (2 to 5 years) behavioral characteristics such as impulsivity, noncompliance, disobedience, and aggression are seen as early analogs of delinquent behavior. At these young ages, self-report instruments are not practical. Rather, researchers have measured these key indicators through either parental reports or observational ratings. Many studies of youngsters at these ages have used Achenbach's (1992) Child Behavior Checklist (CBCL), which is a parent-completed inventory and has versions for children as young as 2 to assess "externalizing" problem behaviors. Studies using either the CBCL, some other parental or teacher report of problem behaviors, or observational ratings have demonstrated that there is a relationship between these early manifestations of problem behavior and antisocial behavior in school-age children (Belsky et al., 1996; Campbell, 1987; Richman et al., 1982; Shaw and Bell, 1993).[2]

Starting at school age, the range of antisocial behaviors expands to include stubbornness, lying, bullying, and other externalizing problems (Loeber et al., 1993). School-age children, even those as young as first grade, begin to participate in delinquent behaviors. However, self-report instruments of delinquent behavior have rarely been administered to preteenage children (Loeber et al., 1989). Some studies have administered self-report instruments to 10 or 11 year olds, slightly modifying the standard delinquency items (Elliott et al., 1985).

Loeber et al. (1989) provide one of the few attempts not only to gather self-report information from children under 10 but also to examine the reliability of those reports. They surveyed a sample of 849 first-grade and 868 fourth-grade boys using a 33-item self-reported antisocial behavior scale. This is a younger-age version of the self-reported delinquency index used by the three projects of the Program of Research on the Causes and Correlates of Delinquency. Items that were age appropriate were selected, and some behaviors were placed in a number of different contexts in order to make them less abstract for the younger children. A special effort was made to ensure that each child understood the question by preceding each behavior with a series of questions to ascertain whether the respondent knew the meaning of the behavior. If the child did not understand the question, the interviewer gave an example and then asked the child to do the same. If the child still did not understand the question, the item was skipped. The parents and teachers of these children also were surveyed using a combination of the appropriate CBCL and delinquency items.

Loeber and colleagues reported that the great majority of boys understood most of the items. First-grade boys did have problems understanding the items regarding marijuana use and sniffing glue, and fourth-grade boys had difficulty understanding the question regarding sniffing glue.

---

[2]The CBCL also assesses internalizing problems.

To assess the validity of self-reported delinquent behavior among elementary school children, Loeber and his colleagues compared the children's self-reports to parental reports about similar behaviors. They found a surprisingly high degree of concordance between children's and parents' reports about the prevalence of delinquent behavior. This is especially true for behaviors that are likely to come to the attention of parents, such as aggressive behaviors and school suspension. Concordance was higher for first graders than fourth graders, which Loeber et al. suggest would be expected since parents would be more likely to know about any misbehavior that takes place at younger ages. These findings are encouraging and suggest that self-report instruments, if administered with concern for the age of the respondents, can be used for very young children.

Self-Report Measures for Adults

Interest in assessing antisocial behavior across the life span has also led to an increasing number of longitudinal surveys that have followed respondents from their adolescent years into early adulthood (e.g., Elliott, 1994; Farrington, 1989b; Hawkins et al., 1992; Huizinga et al., 1998; LeBlanc, 1989; Loeber et al., 1998; Krohn et al., 1997). The concern in constructing self-report instruments for adults is to include items that take into account the different contexts in which crime occurs at these ages (e.g., work instead of school), the opportunities for different types of offenses (e.g., domestic violence, fraud), the inappropriateness or inapplicability of offenses that appear on adolescent self-report instruments (e.g., status offenses), and the potential for very serious criminal behaviors, at least among a small subset of chronic violent offenders.

Weitekamp (1989) has criticized self-report studies for not only being predominantly concerned with the adolescent years but also, when covering the adult years, for using the same items used for juveniles. He argues that even such studies as the NYS (Elliott, 1994) do not include many items that are more serious, and therefore appropriate for adults, than the items included in the original Short and Nye study (1957). Weitekamp asserts that different instruments need to be used during different life stages. Doing so, however, raises questions about construct continuity. If researchers want to document the change in the propensity to engage in antisocial behavior throughout the life course, it must be assumed that different items used to measure antisocial behavior at different ages do indeed measure the same underlying construct. LeBlanc (1989) suggests that a strategy of including different but overlapping items on instruments covering different ages across the life span is the best compromise.

There have been relatively few assessments of the validity of self-report data collected from adults. The data that are available suggest that the validity of adult self-report data is not fundamentally different from that of adolescent self-report data, however. For example, the validity data from Maxfield and his colleagues (2000) and from the DUF project presented above are from adult samples. Their estimates of validity are in the same range as those of most adolescent surveys. Elliott (1994) has presented information from the NYS that suggests adult self-report data are more congruent with adult arrests than juvenile self-report data are with juvenile arrests.

Panel or Testing Effects

Developments in self-report methods have improved the quality of data collected and have expanded their applicability to the study of antisocial behavior throughout the life course. While these advances are significant, they have increased the potential for the data to be contaminated by testing or panel effects. Testing effects are any alterations of a respondent's response to an item or scale that is caused by the prior administration of the same item or scale (Thornberry, 1989).

Improvements in self-report instruments have led to the inclusion of a longer list of items in order to tap more serious offenses, and often a number of follow-up questions are asked. The more acts that a respondent admits to, the longer the overall interview will take. The concern is that this approach will make respondents increasingly unwilling to admit to delinquent acts because those responses will increase the overall length of the interview. This effect likely would be unequally distributed across respondents because those who had the most extensive involvement in delinquency would have the most time to lose by answering affirmatively to the delinquency items.

It is also possible that the simple fact that a respondent is reinterviewed may create a generalized fatigue and lead to decreased willingness by the respondent to respond to self-report items. Research using the National Crime Victimization Survey found that the reduction in reporting was due more to the number of prior interviews than to the number of victimizations reported in prior interviews (Lehnen and Reiss, 1978).

Three studies have examined testing effects in the use of self-report studies; all are based on data from the NYS (Elliott et al., 1985). They were conducted by Thornberry (1989), Menard and Elliott (1993), and Lauritsen (1998). The NYS surveyed a nationally representative sample of 1,725 youth ages 11 to 17 in 1976. The same subjects were reinterviewed annually through 1981. These data allow researchers to examine age-specific prevalence rates by the number of times a respondent was interviewed. For example, some respondents were 14 at the time of their first interview; some were 14 at their second interview (the original 13-year-old cohort); some were 14 at their third interview (the original 12-year-old cohort); and so forth. Because of this, a 14-year-old prevalence rate can be calculated from data collected when the respondents were interviewed for only the first time, from data collected when they were interviewed a second time, etc. If a testing or panel effect plays a role in response rates, the more frequently respondents are interviewed the lower the age-specific rates should be.

Thornberry (1989) analyzed these rates for 17 NYS self-report items representing the major domains of delinquency and, for the most part, the most frequently occurring items. The overall trend suggests a panel effect. For all offenses except marijuana use, comparisons between adjacent waves indicated that age-specific prevalence rates decreased more often than they increased. For example, comparing the rate of gang fights from wave to wave, Thornberry found that for 67 percent of the comparisons there was a decrease in age-specific prevalence rates, whereas there was an increase in only 20 percent of the comparisons and in 13 percent there was no change. The magnitude of the changes was substantial in many cases. For example, for stealing something worth $5 to $50, the rate for 15-year-olds dropped by 50 percent for 15-year-olds from wave 1 to wave 4.

The NYS did not introduce detailed follow-up questions to the delinquency items until the fourth wave of data collection. The data analyzed by Thornberry show that the decline in reporting occurred across all waves. Hence, it appears that the panel design itself, rather than the design of the specific questions, had the effect of decreasing prevalence rates. The observed decline in age-specific rates could be due to an underlying secular drop in offenses during these years (1976-1981). Cross-sectional trend data from the Monitoring the Future (MTF) study, which cannot be influenced by a testing effect, do not indicate any such secular decline (see Thornberry, 1989).

Menard and Elliott (1993) reexamined this issue using both NYS and MTF data. They rightfully pointed out that comparisons between these studies need to be undertaken cautiously because of differences in samples, design features, item wording, and similar concerns. Menard and Elliott's analysis also showed that at the item level, declining trends are more evident in the NYS data than the MTF data. Most of these year-to-year changes are not statistically significant, however. Menard and Elliott then used a modified Cox-Stuart trend test to examine short-term trends in delinquency and drug use. Overall, the trends for 81 percent of the NYS offenses are not statistically significant and about half of the MTF trends are. But an examination of the trends for the 16 items included in their Table 2 indicates that there are more declining trends in the NYS data, 9 of 16 for the 1976-1980 comparisons and 7 of 16 for the 1976-1983 comparisons, than there are for the MTF data, 3 of 16 in both cases. Menard and Elliott focus on the statistically significant effects, which do indicate fewer declining trends in the NYS than is evident when one focuses on all trends, regardless of the magnitude of the change.

More recently, Lauritsen (1998) examined this topic using hierarchical linear models to estimate growth curve models for general delinquency and serious delinquency. She limited her analysis to four of the seven cohorts in the NYS, those who were 11, 13, 15, and 17 years old at wave 1. For those who were 13, 15, or 17 at the start of the NYS, involvement in both general delinquency and serious delinquency decreased significantly over the next four years. For the 11-year-old cohort, the rate of change was also negative but not statistically significant. This downward trajectory in the rate of delinquent behavior for all age cohorts is not consistent with theoretical expectations or with what is generally known about the age-crime curve. Also, as Lauritsen points out, it is not consistent with other data on secular trends for the same time period (see also Thornberry, 1989; Osgood et al., 1989).

Finally, Lauritsen examined whether this testing effect is due to the introduction of detailed follow-up questions during wave 4 of the NYS or whether it appeared to be produced by general panel fatigue. Her analysis of individual growth trajectories indicates that the decline is observed across all waves. Thus she concludes, as Thornberry did, that the reduced reporting is unlikely to have been produced by the addition of follow-up questions.

Overall, Lauritsen offers two explanations for the observed testing effects. One concerns generalized panel fatigue, suggesting that as respondents are asked the same inventory at repeated surveys they become less willing to respond affirmatively to screening questions. The second explanation concerns a maturation effect in which there is change in the content validity of the self-report questions with age. For example, how respondents interpret a question on simple assault and the type of behavior they consider relevant for responding to the question may be quite different for 11 and 17 year olds. This would not account for the drop in the age-specific rates observed by Thornberry (1989), however.

The studies by Thornberry and Lauritsen suggest that it is likely there is some degree of panel bias in self-report data collected in longitudinal panel studies. The analysis by Menard and Elliott indicates that this is indeed just a suggestion at this point, as the necessary comparisons between panel studies and cross-sectional trend studies are severely hampered by the lack of comparability in item wording, administration, and other methodological differences. Also, if there are testing effects, neither Thornberry nor Lauritsen argues that they are unique to the NYS. It just so happens that the sequential cohort design of the NYS makes it a good vehicle for examining this issue. The presumption, unfortunately, is that if testing effects interfere with the validity of the NYS data, they also interfere with the validity of other longitudinal data containing self-report information. This is obviously a serious matter because etiological research has focused almost exclusively on longitudinal designs during the past 20 years. Additional research to identify the extensiveness of testing effects, their sources, and ways to remedy them are certainly a high priority.

Validity of Self-Reports Across Developmental Stages

Earlier we reviewed the literature that assessed the criterion validity of self-report data. Almost all of those studies assess criterion validity at a single point in time. There has been little systematic investigation of validity at different ages, especially for the same subjects followed over time. Because of that, we have begun to assess this issue using the self-report and official data collected in the Rochester Youth Development Study. As in previous studies, two comparisons can be made: (1) the prevalence of self-reported arrests versus the prevalence of official arrests and (2) the prevalence of self-reported delinquency and drug use versus the prevalence of official arrests. We combine the delinquency and drug use items into one self-report inventory since youth can be, and are, arrested for this full range of illegal behaviors. We expect, of course, positive correlations across these alternative measures of involvement in crime.

Table 3-1 presents the results for the total Rochester sample at each of

TABLE 3-1 Yule's Q Comparing the Prevalence of Self-Reported and Official Data, Rochester Youth Development Study, Total Panel

	Wave											
	2	3	4	5	6	7	8	9	10	11	12	Mean
Self-reported arrests[a] with official arrests (n = 834 to 940)	0.83	0.89	0.79	0.86	0.80	0.68	0.78	0.80	0.80	0.86	0.84	0.81
Self-reported delinquency/ drug use with official arrests (n = 836 to 943)	0.48	0.64	0.61	0.57	0.50	0.41	0.52	0.44	0.44	0.45	0.45	0.50

NOTE: The self-report data and the official data for waves 10, 11, and 12 are annual rates. The data for waves 2 through 9 cover six-month periods.

[a] The method of asking about arrests changed during the course of the study. In waves 2 and 3, respondents were asked if they had been arrested or picked up by the police in the last six months. In waves 4 through 11, the arrest questions were presented as follow-up questions to the self-reported delinquency/drug use inventory and a global arrest question (arrested or picked up for anything else) was included at the end. In wave 12 only the global question was asked.

11 waves of data, waves 2 through 12. This allows us to assess criterion validity from early adolescence (the average age at wave 2 is 14) to early adulthood (the average age at wave 12 is 22). The self-reported arrest measure asks respondents if they had been arrested or picked up by the police since the last interview. The self-reported delinquency data are for our general delinquency index, which includes a variety of offenses from trivial to serious. The official arrest file contains information on arrests and official warnings during the juvenile years and arrests during the adult years. Rochester city, Monroe County, and New York state files were searched. Each arrest was assigned to an interview wave.

There is a high degree of concordance between the official arrest histories and the self-reported arrest histories for the Rochester subjects. We use Yule's Q, a standard measure of association for two-by-two contingency tables, that varies from 0 to 1 (Christensen, 1997). The average Yule's Q is 0.81 across the 11 waves, and the range is from 0.68 to 0.89. Subjects who have an official contact or arrest were, generally speaking, willing to report that to their interviewers. There does not appear to be a strong developmental trend in the validity of these data.

The second panel in Table 3-1 presents the association between official arrests and self-reported general delinquency and drug use. If the self-report data are valid, it can be expected that subjects who report offending will be more apt to have an official record than subjects who do not. This is generally what we see, although consistent with the literature, these coefficients are somewhat lower than those in the top panel.

The average Yule's Q across the 11 waves is 0.50, with a range between 0.41 and 0.64. Here there does seem to be a slight downward drift in the size of the relationship over time. During the first few waves, the correlations are in the 0.5 to 0.6 range, but by the last four waves they are in the 0.40 to 0.45 range. The coefficients for the early waves are similar to those reported in previous studies of adolescents (e.g., Hindelang et al., 1981).

It is not yet clear why these coefficients decline over time. The drop in the validity estimates for self-reported delinquency is consistent with a testing effect, although the major decline does not occur until the last few waves. The absence of a strong trend in the self-reported arrest data argues against a testing effect, however, since for most waves these questions were embedded in the self-report follow-up questions. An alternative explanation concerns the changing nature of criminal behavior. It is possible that offenses committed at these ages (early 20s) are less public and therefore somewhat less well correlated with arrest data.

A major question about the validity of self-report data concerns differential levels of reporting by race/ethnicity and gender. Table 3-2 presents comparisons for male and female respondents separately. Overall, there is a somewhat higher degree of validity for the female respondents than the males. The average Yule's Q for the comparison between official arrests and self-reported arrests is 0.74 for males and 0.84 for females. There is no evidence of a strong developmental trend for these data. For the comparison between self-reported delinquency/drug use and official arrests, the average association is 0.43 for the males and 0.52 for the females. For the male respondents, the size of the coefficients tails off somewhat at the older ages. The results for females are unstable, probably because of the low number of females who were arrested at these six-month intervals.

Table 3-3 presents the results by race/ethnicity. When attention focuses on the association between self-reported arrests and official arrests, there is no evidence of differential validity. The mean Yule's Q for African Americans is 0.82, for Hispanics 0.80, and for whites 0.83. There are no strong developmental trends across time for any of the three groups.

The comparison between self-reported delinquency/drug use and official arrests is hampered by our inability to estimate Yule's Q for the white subjects. At 9 of the 11 waves there are empty cells and/or expected cell frequencies of less than 5. Nevertheless, there does seem to be some evidence of differential validity across racial groups. The mean for African Americans is 0.47; for Hispanics, 0.67.

Overall, when self-reported arrests and official arrests are compared, there is little evidence of differential attrition by gender or race/ethnicity and all the coefficients are reasonably high. For the comparison between self-reported delinquency/drug use and official arrests, however, validity is lower for African Americans than Hispanics. This finding is consistent with previous research and must be taken into account when using self-report data.

Similarity of Results

In the past quarter century criminological research has increasingly relied on longitudinal studies to describe and explain patterns of criminal behavior. Much of this research, especially the descriptive studies, has used official measures of crime, but there has been growing use of self-report data, especially in etiological studies. An important but understudied topic is the extent to which these two measures provide the same or different results with respect to key criminal career parameters.

Farrington and colleagues (2000) have begun to address this issue with data from the Seattle Social Development Project. Focusing on the juvenile years, ages 11 to 17, they compared results based on self-reports to those based on court referrals. There was a good deal of similarity across the methods. In particular, similar patterns were found for variations in prevalence by age, the level of continuity in commission of offenses, and the relationship between age of onset and later frequency of committing offenses. There were also some notable differences. "In self-reports, prevalence and individual offending frequency were higher, the age of onset was earlier, and the concentration of offending was greater" (Farrington et al., 2000:21). Also, there was less variation in the individual offending rate by age for the official data compared to self-reports.

While this study is a good first step to take in exploring the issue, it is not yet clear whether the glass is half empty or half full. Additional investigation is needed to identify which criminal career parameters are similar and which are different, across a variety of data sets.

Longitudinal research has demonstrated a substantial degree of continuity in offending. Past offending is related to future offending in both official and self-report data (Farrington et al., 2000). A current controversy in the criminological literature is the source of this continuity. Some argue that it is generated by static, time-stable characteristics (persistent population heterogeneity); others argue that it is generated by dynamic, time-varying processes (state dependence). A number of studies have empirically examined whether the association between past and future offending persists after stable individual differences are taken into account (Nagin and Paternoster, 1991, 2000). If it does, we assume the association is due in part to dynamic processes. Previous studies have used both self-reported data and official data, but typically not on the same individuals. Unfortunately, the results vary somewhat by type of data. Studies based on self-reports are more apt to find a state dependence effect than are studies based on official data.

To examine this issue more systematically, Brame, Bushway, Paternoster, and Thornberry (2001) used both self-report data and official data on subjects in the Rochester Youth Development Study. Separate models were estimated for violent and property offenses, for self-report and official data, and for the younger (≤ 13) and older (≥ 14) groups in the Rochester sample, yielding a total of eight models.

TABLE 3-2 Yule's Q Comparing the Prevalence of Self-Reported and Official Data, Rochester Youth Development Study, by Gender

	Wave											
	2	3	4	5	6	7	8	9	10	11	12	Mean
Male												
Self-Reported Arrests with Official Arrests (n = 597 to 683)	0.84	0.84	0.73	0.83	0.61	0.67	0.68	0.75	0.66	0.80	0.77	0.74
Self-Reported Delinquency/ Drug Use with Official Arrests (n = 599 to 686)	0.44	0.62	0.45	0.49	0.33	0.56	0.28	0.53	0.30	0.43	0.27	0.43

Female										
Self-Reported Arrests with Official Arrests (n = 234 to 257)	0.80	0.93	0.85	0.88	0.95	0.49	0.89	0.76	0.92	0.92	0.89	0.84
Self-Reported Delinquency/ Drug Use with Official Arrests (n = 234 to 257)	0.54	0.66	0.87	0.65	0.69	0.00	0.84	0.08	0.52	0.37	0.55	0.52

NOTE: The Yule's Q values in Table 3-1 for the total sample are not simple arithmetic means of the comparable Yule's Q values by gender because of the sample weights that are applied.

TABLE 3-3 Yule's Q Comparing the Prevalence of Self-Reported and Official Data, Rochester Youth Development Study, by Race/Ethnicity

	Wave											
	2	3	4	5	6	7	8	9	10	11	12	Mean
African American Self-Reported Arrests with Official Arrests (n = 324 to 383)	0.85	0.92	0.78	0.85	0.79	0.63	0.78	0.80	0.82	0.90	0.88	0.82
Self-Reported Delinquency/ Drug Use with Official Arrests (n = 326 to 385)	0.44	0.54	0.54	0.58	0.39	0.35	0.57	0.26	0.45	0.45	0.56	0.47

Hispanic											
Self-Reported Arrests with Official Arrests (n = 104 to 115)	0.70	0.71	0.83	0.86	0.88	0.75	0.83	0.86	0.90	0.76	0.80
Self-Reported Delinquency/ Drug Use with Official Arrests (n = 104 to 115)	0.82	0.79	0.86	0.43	0.66	0.87	0.55	0.77	0.60	0.64	0.35	0.67
White[a]											
Self-Reported Arrests with Official Arrests (n = 158 to 174)	0.87	0.88	0.73	0.85	0.97	0.83	0.76	0.81	0.70	0.78	0.91	0.83

[a]The associations between self-reported delinquency/drug use and official arrests are not reported for white subjects due to empty cells and/or expected cell frequencies of less than 5 in 9 of the 11 waves.

In all but one case (violent offenses for the older group measured by official data) there is a positive effect of past offenses on future offenses after unobserved heterogeneity is held constant. In the exceptional case the number of arrests for violent offenses is so sparse over time that we do not think the estimate is reliable.

Overall, therefore, these results based on the same subjects suggest that self-report and official data yield the same substantive conclusion on this central issue. Both data sources indicate there are both static and dynamic processes at work that produce the observed association between past and future offenses.

CONCLUSIONS

The self-report method for measuring crime and delinquency has developed substantially since it was introduced a half century ago. It is now one of the fundamental ways to scientifically measure criminality, and it forms the bedrock of etiological studies. The challenges confronting this approach to measurement are daunting; after all, individuals are asked to tell about their own undetected criminality. Despite this fundamental challenge, the technique seems to be successful and capable of producing valid and reliable data.

Early self-report scales had substantial weaknesses, containing few items and producing an assessment of only minor forms of offending. Gradually, as the underlying validity of the approach became evident, the scales expanded in terms of breadth, seriousness, and comprehensiveness. Contemporary measures typically cover a wide portion of the behavioral domain included under the construct of crime and delinquency. These scales are able to measure serious as well as minor forms of crime, major subdomains (such as violence, property crimes, and drug use), and different parameters of criminal careers (such as prevalence, frequency, and seriousness) and identify high-rate as well as low-rate offenders. This is substantial progress for a measurement approach that began with a half dozen items and a four-category response set.

The self-report approach to measuring crime has acceptable, albeit far from perfect, reliability and validity. Of these two basic psychometric properties, the evidence for reliability is stronger. There are no fundamental challenges to the reliability of these data. Test-retest measures (and internal consistency measures) indicate that self-reported measures of delinquency are as reliable as, if not more reliable than, most social science measures.

Validity, as noted above, is much harder to assess as there is no gold standard by which to judge self-reports. Nevertheless, current scales seem to have acceptable levels of content and construct validity. The evidence for criterion validity is less clear-cut. At an overall level, criterion validity seems to be in the moderate-to-strong range. While there is certainly room for improvement, the validity appears acceptable for most analytical tasks. At a more specific level, however, there is a potentially serious problem with differential validity in that African American males have lower validity than do Hispanic males. Additional research on this topic is imperative.

While basic self-report surveys appear to be reliable and valid, researchers have experimented with a variety of data collection methods to improve the quality of reporting. Several of these attempts have produced ambiguous results; for example, there is no clear-cut benefit to the mode of administration (interview vs. questionnaire) or the use of randomized response techniques. There is one approach that appears to hold great promise—audio-assisted computerized interviews, which produce increased reporting of many sensitive topics, including delinquency and drug use. Greater use of this approach is warranted.

In the end, the available data indicate that the self-report method is an important and useful way to collect information about criminal behavior. The skepticism of early critics like Nettler (1978) and Gibbons (1979) has not been realized. Nevertheless, the self-report technique can clearly be improved. The final topic addressed in this chapter concerns suggestions for future research.

Future Directions

Much of our research on reliability and validity simply assesses these characteristics; there is much less research on improving their levels. For example, it is likely that both validity and reliability would be improved if we experimented with alternative items for measuring the same behavior and identified the strongest ones. Similarly, reliability and validity vary across subscales (e.g., Huizinga and Elliott, 1986); improving subscales will not only help them but also the overall scale as they are aggregated.

This chapter raised the issue of differential validity for African American males. It is crucial that more is learned about the magnitude of this bias and, if it exists, its source. Future research should address this issue directly and attempt to identify techniques for eliminating it. These research efforts should not lose sight of the fact that the problem may be with the criterion variable (official records) and not the self-reports.

The self-report method was developed in and for cross-sectional studies. Using it in longitudinal studies, especially ones that cover major portions of the life course, creates a new set of challenges. Maintaining the age appropriateness of the items while at the same time ensuring content validity is a knotty problem that we have just begun to address. There is some evidence that repeated measures may create testing effects. More research is needed to measure the size of this effect and its sources and to identify methods to reduce its threat to the validity of self-report data in the longitudinal studies so crucial to etiological investigation.

The similarities and differences in our understanding of criminal career parameters in self-report data and official data are just beginning to be investigated. This approach began with official data but is increasingly coming to rely on self-report data. It is important that we understand more about the validity of both types of data for these purposes.

Finally, we recommend that methodological studies be done in a crosscutting fashion so that several of these issues—reliability and validity, improved item selection, assessing panel bias—can be investigated simultaneously. In particular it is important to examine all of these methodological issues when data are collected using audio-assisted computerized interviewing. For example, studies that have found differential validity or testing effects have all used paper-and-pencil interviews. Whether these same problems are evident under the enhanced confidentiality of audio interviews is an open question. It is clearly a high-priority one as well.

There is no dearth of work that can be done to assess and improve the self-report method. If the progress of the past half century is any guide, we are optimistic that the necessary studies will be conducted and that they will improve this basic way of collecting data on criminal behavior.

REFERENCES

Achenbach, T.M.
    1992 *Manual for the Child Behavior Checklist/2-3 and 1992 Profile.* Burlington: University of Vermont.

Akers, R.L.
    1964 Socio-economic status and delinquent behavior: A retest. *Journal of Research in Crime and Delinquency* 1:38-46.

Akers, R.L., M.D. Krohn, L. Lanza-Kaduce, and M. Radosevich
 1979 Social learning and deviant behavior: A specific test of a general theory. *American Sociological Review* 44:636-655.
Akers, R.L., J. Massey, W. Clarke, and R.M. Lauer
 1983 Are self-reports of adolescent deviance valid? Biochemical measures, randomized response, and the bogue pipeline in smoking behavior. *Social Forces* 62(September):234-251.
Anderson, L.S., T.G. Chiricos, and G.P. Waldo
 1977 Formal and informal sanctions: A comparison of deterrent effects. *Social Problems* 25:103-112.
Aquilino, W.S.
 1994 Interview mode effects in surveys of drug and alcohol use. *Public Opinion Quarterly* 58:210-240.
Aquilino, W.S., and L. LoSciuto
 1990 Effects of interview mode on self-reported drug use. *Public Opinion Quarterly* 54:362-395.
Beebe, T.J., P.A. Harrison, J.A. McRae, Jr., R.E. Anderson and J.A. Fulkerson
 1998 An evaluation of computer-assisted self-interviews in a school setting. *Public Opinion Quarterly* 62:623-632.
Belsky, J., S. Woodworth, and K. Crnic
 1996 Troubled family interaction during toddlerhood. *Development and Psychopathology* 8:477-495.
Belson, W.A.
 1968 The extent of stealing by London boys and some of its origins. *Advancement of Science* 25:171-184.
Blackmore, J.
 1974 The relationship between self reported delinquency and official convictions amongst adolescent boys. *British Journal of Criminology* 14:172-176.
Blumstein, A., J. Cohen, J.A. Roth, and C.A. Visher
 1986 *Criminal Careers and Career Criminals.* Washington, DC: National Academy Press.
Brame, R., S. Bushway, R. Paternoster, and T.P. Thornberry
 2001 Temporal Linkages in Violent and Nonviolent Criminal Activity. Unpublished manuscript, University of South Carolina, Columbia.
Braukmann, C.J., K.A. Kirigin, and M.M. Wolf
 1979 Social Learning and Social Control Perspectives in Group Home Delinquency Treatment Research. Paper presented to the American Society of Criminology, Philadelphia.
Broder, P.K., and J. Zimmerman
 1978 *Establishing the Reliability of Self-Reported Delinquency Data.* Williamsburg, VA: National Center for State Courts.
Browning, K., D. Huizinga, R. Loeber, and T.P. Thornberry
 1999 *Causes and Correlates of Delinquency Program,* Fact Sheet, Washington, DC: U.S. Department of Justice, Office of Juvenile Justice and Delinquency Prevention.
Campbell, S.B.
 1987 Parent-referred problem three-year-olds: Developmental changes in symptoms. *Journal of Child Psychology and Psychiatry* 28:835-845.

Campbell, S.B.
 1990 *Behavioral Problems in Preschool Children: Clinical and Developmental Issues.* New York: Guilford Press.
Christensen, R.
 1997 *Log-Linear Models and Logistic Regression,* 2nd ed. New York: Springer-Verlag.
Clark, J.P., and L.L. Tifft
 1966 Polygraph and interview validation of self-reported delinquent behavior. *American Sociological Review* 31:516-523.
Clark, J.P., and E.P. Wenninger
 1962 Socioeconomic class and area as correlates of illegal behavior among juveniles. *American Sociological Review* 28:826-834.
Conger, R.
 1976 Social control and social learning models of delinquency: A synthesis. *Criminology* 14:17-40.
Drug Use Forecasting
 1990 *Drug Use Forecasting Annual Report: Drugs and Crime in America.* Washington, DC: U.S. Department of Justice.
Dentler, R.A., and L.J. Monroe
 1961 Social correlates of early adolescent theft. *American Sociological Review* 26:733-743.
Elliott, D.S.
 1966 Delinquency school attendance and dropout. *Social Problems* 13:306-318.
 1994 Serious violent offenders: Onset, developmental course, and termination. *Criminology* 32:1-21.
Elliott, D.S., and S.S. Ageton
 1980 Reconciling race and class differences in self-reported and official estimates of delinquency. *American Sociological Review* 45:95-110.
Elliott, D.S., and H.L. Voss
 1974 *Delinquency and Dropout.* Lexington, MA: D.C. Heath.
Elliott, D.S., D. Huizinga, and S.S. Ageton
 1985 *Explaining Delinquency and Drug Use.* Beverly Hills, CA: Sage.
Empey, L.T. ,and M. Erickson
 1966 Hidden delinquency and social status. *Social Forces* 44(June):546-554.
Erickson, M. and L.T. Empey
 1963 Court records, undetected delinquency and decision-making. *Journal of Criminal Law, Criminology, and Police Science* 54:456-469.
Farrington, D.P.
 1973 Self-reports of deviant behavior: Predictive and stable? *Journal of Criminal Law and Criminology* 64:99-110.
 1986 Stepping Stones to Adult Criminal Careers. In *Development of Antisocial and Prosocial Behavior,* D. Olweus, J. Block, and M.R. Yarrow, eds. New York: Academic Press.
 1989a Early predictors of adolescent aggression and adult violence. In *Violence and Victims.* Washington, DC: Springer.
 1989b Self-reported and official offending from adolescence to adulthood. In *Cross-*

National Research in Self-Reported Crime and Delinquency, M.W. Klein, ed. Los Angeles: Kluwer Academic Publishers.

Farrington, D.P., R. Loeber, M. Stouthamer-Loeber, W.B. Van Kammen, and L. Schmidt
 1996 Self-reported delinquency and a combined delinquency seriousness scale based on boys, mothers, and teachers: Concurrent and predictive validity for African-American and Caucasians. *Criminology* 34:493-517.

Farrington, D.P., D. Jolliffe, J.D. Hawkins, R.F. Catalano, K.G. Hill, and R. Kosterman
 2000 Comparing Delinquency Careers in Court Records and Self-reports. Unpublished manuscript, Cambridge University, United Kingdom.

Gibbons, D.C.
 1979 *The Criminological Enterprise: Theories and Perspectives*. Englewood Cliffs, NJ: Prentice-Hall.

Gold, M.
 1966 Undetected delinquent behavior. *Journal of Research in Crime and Delinquency* 3:27-46.
 1970 *Delinquent Behavior in an American City*. Belmont, CA: Brooks/Cole.

Gottfredson, M.R., and T. Hirschi
 1990 *A General Theory of Crime*. Stanford, CA: Stanford University Press.

Hardt, R.H., and S. Petersen-Hardt
 1977 On determining the quality of the delinquency self-report method. *Journal of Research in Crime and Delinquency* 14:247-261.

Hathaway, R.S., E.D. Monachesi, and L.A. Young
 1960 Delinquency rates and personality. *Journal of Criminal Law, Criminology, and Police Science* 50:433-440.

Hawkins, J.D., R.F. Catalano, and J.Y. Miller
 1992 Risk and protective factors for alcohol and other drug problems in adolescence and early adulthood: Implications for substance abuse prevention. *Psychological Bulletin* 112:64-105.

Hepburn, J.R.
 1976 Testing alternative models of delinquency causation. *Journal of Criminal Law and Criminology* 67:450-460.

Hindelang, M.J.
 1973 Causes of delinquency: A partial replication and extension. *Social Problems* 20:471-487.

Hindelang, M.J., T. Hirschi, and J.G. Weis
 1979 Correlates of delinquency: The illusion of discrepancy between self-report and official measures. *American Sociological Review* 44:995-1014.
 1981 *Measuring Delinquency*. Beverly Hills, CA: Sage.

Hirschi, T.
 1969 *Causes of Delinquency*. Berkeley: University of California Press.

Huesmann, L.R., L.D. Eron, M.M. Lefkowitz, and L.O. Walder
 1984 The stability of aggression over time and generations. *Developmental Psychology* 20:1120-1134.

Huizinga, D., and D.S. Elliott
 1983 *A Preliminary Examination of the Reliability and Validity of the National Youth*

Survey Self-Reported Delinquency Indices. National Youth Survey Project Report 27. Boulder, CO: Behavioral Research Institute.
 1986 Reassessing the reliability and validity of self-report delinquent measures. *Journal of Quantitative Criminology* 2:293-327.
Huizinga, D., A.W. Weiher, S. Menard, R. Espiritu, and F.A. Esbensen
 1998 Some not so boring findings from the Denver Youth Survey. Paper presented at the annual meeting of American Society of Criminology, Washington, D.C.
Jensen, G.F.
 1973 Inner containment and delinquency. *Journal of Criminal Law and Criminology* 64:464-470.
Jensen, G.F., and R. Eve
 1976 Sex differences in delinquency. *Criminology* 13:427-448.
Jensen, G.F., M.L. Erickson, and J.P. Gibbs
 1978 Perceived risk of punishment and self-reported delinquency. *Social Forces* 57:57-78.
Jessor, R.
 1998 *New Perspectives on Adolescent Risk Behavior.* New York: Cambridge University Press.
Jessor, R., J.E. Donovan, and F.M. Costa
 1991 *Beyond Adolescence: Problem Behavior and Young Adult Development.* Cambridge, England: Cambridge University Press.
Johnson, R.E.
 1979 *Juvenile Delinquency and Its Origins.* Cambridge, England: Cambridge University Press.
Johnston, L.D., P.M. O'Malley, and J.G. Bachman
 1996 *National Survey Results on Drug Use from the Monitoring the Future Study, 1975-1995.* Washington, DC: U.S. Government Printing Office.
Kaplan, H.B.
 1972 Toward a general theory of psychosocial deviance: The case of aggressive behavior. *Social Science and Medicine* 6:593-617.
Kelly, D.H.
 1974 Track position and delinquent involvement: A preliminary analysis. *Sociology and Social Research* 58:380-386.
Klein, M.W., ed.
 1989 *Cross-National Research in Self-Reported Crime and Delinquency.* Los Angeles: Kluwer Academic Publishers.
Krohn, M.D., G.P. Waldo, and T.G. Chiricos
 1974 Reported delinquency: A comparison of structured interviews and self-administered check-lists. *Journal of Criminal Law and Criminology* 65:545-553.
Krohn, M.D., A.J. Lizotte, and C.M. Perez
 1997 The interrelationship between substance use and precocious transitions to adult statutes. *Journal of Health and Social Behavior* 38:87-103.
Krohn, M.D., T.P. Thornberry, C. Rivera, and M. LeBlanc
 2001 Later delinquency careers. In *Child Delinquents: Development, Intervention, and Service Needs,* R. Loeber and D.P. Farrington, eds. Thousand Oaks, CA: Sage.

Kulik, J.A., K.B. Stein, and T.R. Sarbin
   1968   Disclosure of delinquent behavior under conditions of anonymity and non-anonymity. *Journal of Consulting and Clinical Psychology* 32:506-509.
Lauritsen, J.L.
   1998   The age-crime debate: Assessing the limits of longitudinal self-report data. *Social Forces* 76:1-29.
LeBlanc, M.
   1989   Designing a self-report instrument for the study of the development of offending from childhood to adulthood: Issues and problems. Pp. 371-398 in *Cross-National Research in Self-Reported Crime and Delinquency*, M.W. Klein, ed. Los Angeles: Kluwer Academic Publishers.
Lehnen, R.G., and A.J. Reiss
   1978   Response effects in the National Crime Survey. *Victimology: An International Journal* 3:110-124.
Loeber, R., M. Stouthamer-Loeber, W.B. Van Kammen, and D.P. Farrington
   1989   Development of a new measure of self-reported antisocial behavior for young children: Prevalence and reliability. Pp. 203-225 in *Cross-National Research in Self-Reported Crime and Delinquency*, M.W. Klein, ed. Los Angeles: Kluwer Academic Publishers.
Loeber, R., P. Wung, K. Keenan, B. Giroux, and M. Stouthamer-Loeber
   1993   Developmental pathways in disruptive child behavior. *Development and Psychopathology* 5:101-133.
Loeber, R., D.P. Farrington, M. Stouthamer-Loeber, T.E. Moffitt, and A. Caspi
   1998   The development of male offending: Key findings from the first decade of the Pittsburgh Youth Study. *Studies on Crime and Crime Prevention* 7:1-31.
Matthews, V.M.
   1968   Differential identification: An empirical note. *Social Problems* 14:376-383.
Maxfield, M.G., B.L. Weiler, and C.S. Widom
   2000   Comparing self-reports and official records of arrests. *Journal of Quantitative Criminology* 16:87-100.
Menard, S. and D.S. Elliott
   1993   Data set comparability and short-term trends in crime and delinquency. *Journal of Criminal Justice* 21:433-445.
Moffitt, T.E.
   1993   Life-course-persistent and adolescence-limited antisocial behavior: A developmental taxonomy. *Psychological Review* 100:674-701.
   1997   Adolescence-limited and life-course-persistent offending: A complementary pair of developmental theories. Pp. 11-54 in *Developmental Theories of Crime and Delinquency, Volume 7: Advances in Criminological Theory*, T.P. Thornberry, ed. New Brunswick, NJ: Transaction Publishers.
Nagin, D.S., and R. Paternoster
   1991   On the relationship of past and future participation in delinquency. *Criminology* 29:163-190.
   2000   Population heterogeneity and state dependence: State of the evidence and directions for future research. *Journal of Quantitative Criminology* 16:117-145.

Nettler, G.
    1978   *Explaining Crime.* New York: McGraw-Hill.
Nye, F.I., J.F. Short, and V.J. Olson
    1958   Socioeconomic status and delinquent behavior. *American Journal of Sociology* 63:381-389.
Olweus, D.
    1979   Stability and aggressive reaction patterns in males: A review. *Psychological Bulletin* 86:852-875.
Osgood, D.W., P. O'Malley, J. Bachman, and L. Johnston
    1989   Time trends and age trends in arrests and self-reported illegal behavior. *Criminology* 27:389-418.
Patterson, G.R.
    1993   Orderly change in a stable world: The antisocial trait as a chimera. *Journal of Consulting and Clinical Psychology* 61:911-919.
Patterson, G.R., and R. Loeber
    1982   The Understanding and Prediction of Delinquent Child Behavior. Research proposal to National Institute of Mental Health from the Oregon Social Learning Center, Eugene.
Polk, K.
    1969   Class strain and rebellion among adolescents. *Social Problems* 17:214-224.
Porterfield, A.L.
    1943   Delinquency and outcome in court and college. *American Journal of Sociology* 49:199-208.
    1946   *Youth in Trouble.* Fort Worth, TX: Leo Potishman Foundation.
Reiss, A.J., Jr., and A.L. Rhodes
    1959   *A Socio-Psychological Study of Adolescent Conformity and Deviation.* Washington, DC: U.S. Office of Education.
    1963   Status deprivation and delinquent behavior. *Sociological Quarterly* 4:135-149.
    1964   An empirical test of differential association theory. *Journal of Research in Crime and Delinquency* 1:5-18.
Richman, N., J. Stevenson, and P.J. Graham
    1982   *Preschool to School: A Behavioural Study.* London: Academic Press.
Rojek, D.G.
    1983   Social status and delinquency: Do self-reports and official reports match? In *Measurement Issues in Criminal Justice,* G.P. Waldo, ed. Beverly Hills, CA: Sage.
Sampson, R.J., and J.H. Laub
    1990   Crime and deviance over the life course: The salience of adult social bonds. *American Sociological Review* 55:609-627.
Saris, W.E.
    1991   *Computer-Assisted Interviewing.* Beverly Hills, CA: Sage.
Sellin, T.
    1931   The basis of a crime index. *Journal of Criminal Law and Criminology* 22:335-356.
Shaw, D.S., and R.Q. Bell
    1993   Developmental theories of parental contributors to antisocial behavior. *Journal of Abnormal Child Psychology* 21:35-49.

Short, J.F.
   1957 Differential association and delinquency. *Social Problems* 4:233-239.
Short, J.F., Jr., and F.I. Nye
   1957 Reported behavior as a criterion of deviant behavior. *Social Problems* 5:207-213.
   1958 Extent of unrecorded juvenile delinquency: Tentative conclusions. *Journal of Criminal Law and Criminology* 49:296-302.
Silberman, M.
   1976 Toward a theory of criminal deterrence. *American Sociological Review* 41:442-461.
Slocum, W.L., and C.L. Stone
   1963 Family culture patterns and delinquent-type behavior. *Marriage and Family Living* 25:202-208.
Skolnick, J.V., C.J. Braukmann, M.M. Bedlington, K.A. Kirigin, and M.M. Wolf
   1981 Parent-youth interaction and delinquency in group homes. *Journal of Abnormal Child Psychology* 9:107-119.
Stanfield, R.
   1966 The interaction of family variables and gang variables in the aetiology of delinquency. *Social Problems* 13:411-417.
Thornberry, T.P.
   1987 Toward an interactional theory of delinquency. *Criminology* 25:863-891.
   1989 Panel effects and the use of self-reported measures of delinquency in longitudinal studies. Pp. 347-369 in *Cross-National Research in Self-Reported Crime and Delinquency*, M.W. Klein, ed. Los Angeles: Kluwer Academic Publishers.
Thornberry, T.P., ed.
   1997 *Developmental Theories of Crime and Delinquency.* New Brunswick, NJ: Transaction Publishers.
Thornberry, T.P., and M.D. Krohn
   2001 The development of delinquency: An interactional perspective. Pp. 289-305 in *Handbook of Law and Social Science: Youth and Justice*, S.O. White, ed. New York: Plenum.
Thornberry, T.P., M.D. Krohn, A.J. Lizotte, C.A. Smith, and K. Tobin
   In   *The Toll of Gang Membership: Gangs and Delinquency in Developmental Perspective.*
   Press New York: Cambridge University Press.
Thrasher, F.
   1927 *The Gang: A Study of 1,313 Gangs in Chicago.* Chicago: University of Chicago Press.
Tourangeau, R. and T.W. Smith
   1996 Asking sensitive questions: The impact of data collection, mode, question format, and question context. *Public Opinion Quarterly* 60:275-304.
Tracy, P.E., and J.A. Fox
   1981 The validity of randomized response for sensitive measurements. *American Sociological Review* 46:187-200.
Tremblay, R.E., C. Japel, D. Perusse, P. McDuff, M. Boivin, M. Zoccolillo, and J. Montplaisir
   1999 The search for the age of 'onset' of physical aggression: Rousseau and Bandura revisited. *Criminal Behaviour and Mental Health* 9:8-23.

Turner, C.F., J.T. Lessler, and J. Devore
    1992   Effects of mode of administration and wording on reporting of drug use. Pp. 177-220 in *Survey Measurement of Drug Use: Methodological Studies*, C.F. Turner, J.T. Lessler, and J.C. Gfroerer, eds. Washington, DC: U.S. Department of Health and Human Services.

Vaz, E.W.
    1966   Self-reported juvenile delinquency and social status. *Canadian Journal of Corrections* 8:20-27.

Voss, H.L.
    1963   Ethnic differentials in delinquency in Honolulu. *Journal of Criminal Law and Criminology* 54:322-327.
    1964   Differential association and reported delinquent behavior: A replication. *Social Problems* 12:78-85.
    1966   Socio-economic status and reported delinquent behavior. *Social Problems* 13:314-324.

Waldo, G.P., and T.G. Chiricos
    1972   Perceived penal sanction and self-reported criminality: A neglected approach to deterrence research. *Social Problems* 19:522-540.

Wallerstein, J.S., and C.J. Wylie
    1947   Our law-abiding law-breakers. *Probation* 25:107-112.

Weis, J.G., and D.V. Van Alstyne
    1979   The Measurement of Delinquency by the Randomized Response Method. Paper presented at the meeting of the American Society of Criminology, Philadelphia.

Weitekamp, E.
    1989   Some problems with the use of self-reports in longitudinal research. Pp. 329-346 in *Cross-National Research in Self-Reported Crime and Delinquency*, M.W. Klein, ed. Los Angeles: Kluwer Academic Publishers.

Widom, C.S.
    1989   Child abuse, neglect, and violent criminal behavior. *Criminology* 27:251-271.

Williams, J.R., and M. Gold
    1972   From delinquent behavior to official delinquency. *Social Problems* 20(2):209-229.

Wolfgang, M.E., R.M. Figlio, and T. Sellin
    1972   *Delinquency in a Birth Cohort*. Chicago: University of Chicago Press.

Wolfgang, M.E., T.P. Thornberry, and R.M. Figlio
    1987   *From Boy to Man, From Delinquency to Crime*. Chicago: University of Chicago Press.

Wright, D.L., W.S. Aquilino and A.J. Supple
    1998   A comparison of computer-assisted and paper-and-pencil self-administered questionnaires in a survey on smoking, alcohol, and drug use. *Public Opinion Quarterly* 62:331-353.

Appendix A

Workshop Agenda

Committee on Law and Justice
Committee on National Statistics
National Research Council

AGENDA

July 24, 2000

9:00-9:15 a.m.	Welcoming Remarks Charles Wellford, *Chair*, Committee on Law and Justice Colin Loftin, Workshop Chair Andrew A. White, *Director*, Committee on National Statistics Carol V. Petrie, *Director*, Committee on Law and Justice

Session One

9:15-9:45 a.m.	*Sensitive Questions in Survey Research* Paper by Roger Tourangeau and Madeline E. McNeeley, Joint Program in Survey Methods, University of Maryland, College Park Presented by Madeline E. McNeeley

9:45-10:15 a.m. Comment
Judith Lessler
Research Triangle Institute

James Walker
University of Wisconsin, Madison

10:15-10:30 a.m. Q&A/Discussion

10:30-10:45 a.m. Break

Session Two

10:45-11:15 a.m. *Small Area Estimation*
T.E. Raghunathan, Survey Methodology Program, Survey Research Center, Institute for Social Research, Ann Arbor

11:15-11:45 a.m. Comment
Elizabeth Stasny, Ohio State University, Columbus

Charles Manski, Northwestern University

11:45 a.m.-
12:00 p.m. Q&A/Discussion

12:00-2:00 p.m. Working Lunch
Charles Wellford, University of Maryland, College Park

Session Three

2:00-2:30 p.m. *Comparison of Self-Report and Official Data*
Terence P. Thornberry, School of Criminal Justice, and Marvin D. Krohn, Department of Sociology, University at Albany, State University of New York, Albany

2:30-3:00 p.m.	Comment David Farrington, Institute of Criminology, Cambridge University, Cambridge Laura Dugan Georgia State University Alfred Blumstein Carnegie Mellon University
3:00-3:15 p.m.	Q&A/Discussion
3:15-3:30 p.m.	Break

Session Four

3:30-4:00 p.m.	*Measuring Rare Events in Small Populations* Richard McCleary, School of Social Ecology, University of California, Irvine
4:00-4:30 p.m.	Comment James Lynch, School of Public Affairs, American University, Washington, DC
4:30-4:45 p.m.	Q&A/Discussion
4:45-5:30 p.m.	General Discussion
5:30 p.m.	Reception and Dinner

Appendix B

List of Workshop Participants

Alfred Blumstein
H. John Heinz III School of Public
  Policy and Management
Carnegie Mellon University

Patrick Clark
Office of Research and Evaluation
National Institute of Justice

Jeanette Covington
Department of Sociology
Rutgers University

Marshall DeBerry
Bureau of Justice Statistics
U.S. Department of Justice

Laura J. Dugan
Georgia State University
Department of Criminal Justice

William F. Eddy
Carnegie Mellon University
Institute for Statistics and Its
  Applications

Jeffrey Fagan
School of Public Health
Columbia University

David P. Farrington
Institute of Criminology
Cambridge University

Steven Fienberg
Department of Statistics
Carnegie Mellon University

Darnell Hawkins
African American Studies
University of Illinois, Chicago

Philip Heymann
Harvard Law School
Harvard University

Sally Hillsman
Office of Research and Evaluation
National Institute of Justice

Marvin D. Krohn
Department of Sociology
University at Albany, State
  University of New York

Candace Kruttschnitt
Department of Sociology
University of Minnesota

Judith Lessler
Statistics, Health, and Social Policy
Research Triangle Institute

Colin Loftin
School of Criminal Justice
State University of New York at Albany

James Lynch
School of Public Affairs
American University

Charles Manski
Department of Economics
Northwestern University

Michael G. Maxfield
School of Criminal Justice
Rutgers University

Richard McCleary
School of Social Ecology
University of California, Irvine

Madeline E. McNeeley
School of Criminology and Criminal Justice
University of Maryland, College Park

John Monahan
School of Law
University of Virginia

Colm O'Muircheartaigh
NORC, Statistics and Method
University of Chicago

Joan Petersilia
School of Social Ecology
University of California, Irvine

Trivellore Raghunathan
Institute for Social Research
University of Michigan

Michael Rand
Bureau of Justice Statistics
U.S. Department of Justice

Linda E. Saltzman
Injury Prevention and Control
Centers for Disease Control and Prevention

Sidney M. Stahl
Health Care Organizations and Social Institutions
National Institute on Aging
National Institute of Health

Elizabeth A. Stasny
Center for Survey Research
Ohio State University

Kate Stith
School of Law
Yale University

Terence P. Thornberry
School of Criminal Justice
University at Albany, State University of New York

Roger Tourangeau
Joint Program in Survey Methods
University of Maryland, College Park

James R. Walker
Department of Economics
University of Wisconsin, Madison

Charles Wellford
Center for Applied Policy Studies & School of Criminology and Criminal Justice
University of Maryland, College Park

Cathy Spatz Widom
Criminal Justice and Psychology
State University of New York at Albany

STAFF

Carol V. Petrie, *Director*
Committee on Law and Justice
National Research Council

Andrew A. White, *Director*
Committee on National Statistics
National Research Council

Nancy A. Crowell
Committee on Law and Justice
National Research Council

Ralph Patterson
Committee on Law and Justice
National Research Council

Brenda McLaughlin
Committee on Law and Justice
National Research Council

COMMITTEE ON LAW AND JUSTICE

The Committee on Law and Justice was created in 1978 to provide a rationale and model for the role of federally sponsored research in anticrime programs. It serves as a national focal point for objective analysis of crime and justice issues to inform national policy. The committee brings the knowledge and tools of the social and behavioral sciences to bear on the development of improved policy, research, and evaluation in the areas of crime prevention, intervention, and control, as well as civil justice. It does so primarily by establishing committees to synthesize, analyze, and evaluate the research from a variety of scientific disciplines that have relevance for criminal and civil justice matters. The members of the committee share a commitment to using basic and applied research to advance prevention research, improve all components of the justice system, develop a broader range of sanctions to sentence offenders; improve criminal justice technology, and expand research in support of new models of justice.